W9-BWW-273

COLONIAL ARCHITECTURE
OF THE MID-ATLANTIC

Other volumes in the
ARCHITECTURAL TREASURES
OF EARLY AMERICA
Series

PROPERTY OF

COLONIAL ARCHITECTURE OF THE MID-ATLANTIC

Vol. IV in the Architectural Treasures of Early America Series

From material originally published as
The White Pine Series of Architectural Monographs
edited by
Russell F. Whitehead and Frank Chouteau Brown

Lisa C. Mullins, Editor

Preface by Roy Underhill,
Master Housewright at Colonial Williamsburg

A Publication of The National Historical Society

DISTRIBUTED BY

THE MAIN STREET PRESS • Pittstown, New Jersey

PROPERTY OF
HIGH POINT PUBLIC LIBRARY
HIGH POINT, NORTH CAROLINA

All rights reserved

Copyright © 1987 by The National Historical Society

No part of this book may be used or reproduced in any manner
whatsoever without written permission except in the case of brief
quotations embodied in critical articles and reviews.

Published by
The National Historical Society
2245 Kohn Road, Box 8200
Harrisburg, Pennsylvania 17105

Distributed by
The Main Street Press
William Case House
Pittstown, New Jersey 08867

Distributed simultaneously in Canada by
Methuen Publications
2330 Midland Avenue
Agincourt, Ontario M1S 1P7

Printed in the United States of America

10   9   8   7   6   5   4   3   2   1

Library of Congress Cataloging-in-Publication Data

Colonial architecture of the Mid-Atlantic.
     (Architectural treasures of early America; 4)
       1. Architecture, Modern — 17th–18th centuries — Middle
Atlantic States.   2. Architecture, Moravian — Pennsylvania
— Bethlehem.   3. Architecture, Dutch — New Jersey.
4. Roofs — Middle Atlantic States.   5. Pain, William,
1730?–1790? Builders companion and workman's assistant.
6. Architecture — Early works to 1800.   7. Bethlehem (Pa.)
— Buildings, structures, etc.   I. Mullins, Lisa C.
II. Underhill, Roy.   III. Series: Architectural treasures
of Early America (Harrisburg, Pa.); 4.
NA717.C65   1987   720'.974   87-14213
ISBN 0-918678-23-4

The original photographs reproduced in this publication are from
the collection of drawings and photographs in "The White Pine
Monograph Series, Collected and Edited by Russell F. Whitehead,
The George P. Lindsay Collection." The collection, part of
the research and reference collections of The American Institute
of Architects, Washington, D.C., was acquired by the Institute
in 1955 from the Whitehead estate, through the cooperation of Mrs.
Russell F. Whitehead, and the generosity of the Weyerhauser
Timber Company, which purchased the collection for presentation
to the Institute. The research and reference collections of the
Institute are available for public use. A written request for such
use is required so that space may be reserved and assistance made
available.

8710031

CONTENTS

BLOOD AND MORTAR

Old homes present many mysteries. As Mr. Embury writes in Chapter 1 of this volume, "it is a curious sidelight upon the knowledge of our ancestors to find that people who could work stone so beautifully as the Dutch had no mortar which was durable when exposed to the weather, and the stone walls were therefore protected by overhanging eaves of wood."

The plot thickens with Clifford Wendehack's essay in Chapter 2. "There is abundant proof that practically all of this early masonry was originally laid in ordinary clay . . . the enormous overhang and sweep of the eaves . . . was the means by which water was diverted from the walls of the buildings and prevented the washing out of the clay joints in the masonry." But, he continues, "assuming that this theory of the overhanging eaves is correct, it is not easily explained why the gable walls were made so extremely flat, in fact, more so than in any other type of early American architecture."

This mystery is a double-bitted axe. First, why would the Dutch, the world-class masters of masonry, not use a proper lime mortar that would withstand the weather? Second, what is the true purpose and origin of this overhang that leaves the sides of a building unprotected? If they were intended to shelter inadequate mortar, why were they also found on wooden buildings? I propose some soft answers to these hard questions.

The Dutch were expert masons and, "thrifty" or not, would hardly have used inadequate mortar. Mr. Wendehack noted the high quality of the New Jersey soil, "stone work laid in red clay mortar was practically impervious to moisture, and many examples of clay joints in window sills are still intact." So the clay was good, and even if lime for proper mortar was scarce, it could be used sparingly on the exterior to further protect the clay from the weather. Mr. Wendehack claims, "at later periods of restoration, most of this stonework was pointed up with lime mortar. . . ." Could this be the original lime mortar seal mistakenly identified as secondary work? — Perhaps.

But now, a digression on lime, mortar and plaster.

Limestone was scarce along the New World coast, but plenty of sea shells were available in natural banks, or in those left by centuries of Indian oyster feasts. Bits of Indian pottery can still be seen in the mortar of Charleston, South Carolina, houses. Even Henry David Thoreau, needing lime to build his cabin at Walden Pond, explored different sources. "I had the previous winter made a small quantity of lime by burning the shells . . . which our river affords . . . I might have got good limestone within a mile or two and burned it myself, if I had cared to do so," he wrote.

"Burning" the shells or limestone is the first step in converting them into lime for building. When burned in alternating layers with wood or coal, the calcium carbonate of the shells or stone releases carbon dioxide, converting the original material to calcium oxide, or "quick-lime." It looks unchanged, just slightly yellower than before, but when mixed with water, or "slaked," the startling transformation to calcium hydrate or slaked lime takes place.

Joseph Moxon's 1778 description of the reaction is beautifully, if under-, stated, "and the Fire in Lime burnt, Asswages not, but lies hid, so that it appears to be cold, but Water excites it again, whereby it Slacks and crumbles into fine Powder." I once supplied a mason with a large barrel of burned shells for him to slake. After keeping it for a few days, he filled the barrel of shells to the top with water. Unimpressed with the apparent lack of results, he walked off to complain, rounding the corner just in time. The delayed explosion from the exothermic reaction lime-plastered four buildings and his parked truck.

Some varieties of impure lime may not be completely slaked and, if used too soon, would expand in the work and cause it to "blow." One old solution was to let it sit for two or three months before using it to make mortar or plaster, but this resulted in weaker bonding. The best remedy was, and is, to beat the mixture so thoroughly and so long that none remains unworked. As one seventeenth-century author put it, "'tis a Maxim among old Masons to their Labourers, that they should dilute the mortar with the Sweat of their Brow."

Sand is the other constituent of mortar. Thoreau was picky about his sand, "I brought over some whiter and cleaner sand from the opposite shore of the pond in a boat." Although many ancient texts give precise formulations for mortar and plaster, lime and sand vary in character too much from place to place. A 1703 builder's guide, wisely suggests that it is better to "be regulated by the judgement of experienced and skillful workmen in each particular country, than by any stated proportions."

If protected from moisture and mechanical damage, plaster and mortar should last forever. (Although none has yet.) Considering the difficulties of matching the character of old plasterwork, many experts contend that you are better off replacing, rather than repairing damaged work. Valuable, damaged plaster can be reconsolidated by experts with experience in new chemical treatments. Often, a plaster wall or ceiling, although still held intact by the intermingled ox hair, will become detached from the lath behind it. If you can get behind the loosened plaster you may be able to reanchor it by applying new plaster. Again, it is a job for expert, on-the-scene consultation. Incidentally, you may wish to microscopically examine some of the hair from your plaster to discover what breed of animals were in your area when the plaster was mixed.

Mortar restoration begins with a simple analysis. Crush a sample of the mortar and mix it with water, dry the sand and look for a local match. Whether you are replacing a missing section or repointing (raking out the joints and tucking in new mortar) a weathered wall, *never* use hard cement in a wall originally laid up in soft mortar. Hard cement will not yield when the softer brick swells. The face of the brick or stone will shatter and fall away. Repointing with hard cement is second only to sandblasting in destroying historic brickwork.

So much for the mortar mystery, now what about the overhanging eaves discussed by Mr. Wendehack? The answer may not lie with the Dutch, but with the other immigrants mentioned earlier in his essay. "Slavery was introduced in New Jersey in 1664, and the land owners were encouraged to increase the number of slaves. . . ." Scholars such as John Michael Vlach now believe that, rather than contributing no more than "unlimited, inexpensive labor," the Africans gave the sweltering Europeans the means to survive the heat and humidity of their new home. Writes Mr. Vlach, "Verandas are common to African house design. Soon after both slaves and their masters arrived in the New World, a cross-cultural encounter occurred, and generations of white builders adopted the custom of porch building." The Dutch, like their neighbors to the south, might well have learned from the African innovations in architectural form.

Rocking chairs, anyone?

ROY UNDERHILL
MASTER HOUSEWRIGHT
COLONIAL WILLIAMSBURG

Farmhouses of
New Netherlands

Text by
Aymar Embury II
Photographs by
Frank Cousins and John Wallace Gillies
Originally published in 1915 as White Pine Monograph
Volume I, Number 3

Front Entrance Detail
VREELAND HOUSE, NORDHOFF, NEW JERSEY
An unusually good example of carpenter carving done with a gouge.

FARMHOUSES OF NEW NETHERLANDS

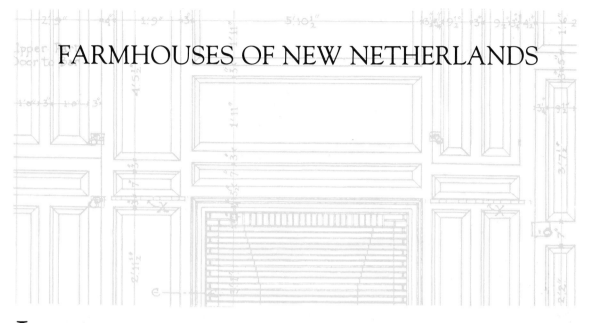

LONG after the Colonial work of New England and the South became well known to the architects, and had become regarded by them as a suitable source from which to draw precedents for modern work, the remaining examples of the work of the Dutch in their colony of New Netherlands remained unnoticed and neglected. It is not easy to discover why this should have been, since much of it is in close proximity to New York City, some of it indeed within the city limits, and these examples are not inferior in charm, less in number, or of a later date than the Colonial work of Massachusetts and Virginia.

The settlement of New Netherlands antedated by some years that of New England, and its development was steady and rapid, the colonists pushing out from New York along the river valleys and Indian trails which formed the natural means of communication in a country where roads were still to be constructed. Many of these early Dutch houses still exist, and although the area in which they occur is comparatively small, it must have been, for a farming community, very thickly populated and extremely prosperous. The age of these houses cannot be determined with any real accuracy, and while the earliest of them appear to have been erected about the same time as the earliest remaining examples in New England or Virginia, the very natural tendency to exaggerate the age of old work has probably been not less apparent in New Netherlands than in New England. The whole question of the dates of old work is a rather delicate one, and I have found in all parts of the American colonies that the dates assigned to old buildings were those at which some por-

tions of them had been built, although the entire building might have been reconstructed since that time.

In selecting the subjects for the illustrations for this article, then, I have been unable to find in many cases any real historic evidence as to the dates of construction, and have been obliged to accept family traditions or the records of the local historic societies as guides, and these dates are offered with reserve. The fact is that in most cases the testimony as to the age is probably no better than that given me by a negro employee on one of the old farms, who told me that the house was built "so dog-gone long ago that there ain't nobody remembers when she was built." I have gone into this question of dates with some particularity, because the determination of the sources and progress of any style must rest primarily upon the comparison of houses in their chronological order, assuming, of course, variances in the style arising from local conditions. Now while this evidence is very far from complete, it is convincing on one point, namely, that the Dutch early found their *métier*, and pursued it substantially unchanged up to, and in some cases even through, the period of the Classic Revival. The difference between the earliest of the Dutch houses and the latest is far less marked than the difference between the early and late houses of New England and the Southern colonies, and without previous knowledge as to the age of the remaining Dutch buildings, it would be practically impossible to pick certain of them out as being the prototypes of the style and others as examples of the style developed.

The most curious thing about the architecture of New Netherlands is that which strikes us in the other colonies, namely, the almost complete renunciation by the colonists of ideals, processes and precedents of their mother country. The Dutch houses in Long Island and New Jersey resembled nothing but themselves, and were even more radically different from the work of the Dutch in Holland than they were from the work of the other colonists. This difference is not alone a question of material, which might be expected in a new country, but is also a question of

SHENKS-CROOK HOUSE—1656—BERGEN BEACH, FLATLANDS, NEW YORK

form and of detail. The steep-pitched roofs of Holland were here transformed into low gentle lines, and the narrow flat cornices of the mother country were replaced by broad overhanging eaves, from which Classic treatment in general was absent. It was an architecture altogether autochthonous, and not the less interesting for that reason.

The characteristics of the Dutch work are by this time fairly well known: the houses are for the most part one story in height, with low curved overhanging eaves on the front and rear, and an almost total suppression of cornices or rake moldings on the gable-ends. The earliest buildings apparently had single pitched roofs; the gambrel form, so common in these colonies that the term "Dutch roof" has become synonymous with "gambrel," was a thing of later development, although toward the latter part of the seventeenth century it already had become customary; but aside from this one change in the roof shape, apparently the only variation from type was the gradual introduction of a piazza or stoop under the overhanging eaves; and this, too, must

LAKE TYSEN HOUSE, NEW DORP, STATEN ISLAND, NEW YORK

have occurred at a very early date.

The materials in the Dutch work were those used in the other colonies: shingles and clapboards, stone and brick for wall covering, and hewn timbers for the frames. These materials were, however, mingled together with much more freedom than we customarily find in the other colonies, and were perhaps treated with a little better realization of the artistic effect possible from careful selection of materials and appropriate treatment of their surfaces than was elsewhere the case. I do not know of any material used in Colonial times which was so beautifully handled as the red sandstone from which the bodies of many of the houses in Bergen and Hudson counties in New Jersey were built. The entrance sides of the houses were invariably better finished than the others, and were usually of coursed ashlar with either fine picked or four cut surfaces, small joints and neatly cut sills. The lintels were flat arches, often of wood and with wooden carved key blocks, painted and sanded to represent stone. The other sides of these buildings were of rougher stone or of wood or of brick, handled with a facility and playfulness which in no way detracted from the dignity and attractiveness of the whole building.

We find the same motive in most of the houses still remaining. Each consists of a central mass with one or two wings, invariably placed on the gable-ends, but it is probable that the original houses were single rectangular blocks which now constitute the central portions or in some cases are now the wings, to which the main bodies of the houses have been added.

The materials varied with the location: in Long Island the exteriors were of wood, generally white pine shingles but sometimes white pine clapboards; in Staten Island and New York they were sometimes of stone whitewashed or stuccoed, and sometimes of shingles, stone apparently having been used where it was not too hard to cut, and wood used elsewhere. In New Jersey, where the fields were covered with erratic glacial drift of red sandstone, and had to be cleared before cultivation, the bodies of the houses up to the second-story line were generally built of this stone, with the gable-ends, roofs and wings of wood. This red sandstone split readily, was easy to work, and hardened upon exposure to the air, and was therefore chosen in many instances; but it is a curious sidelight upon the knowledge of our ancestors to find that people who could work stone so beautifully as the Dutch had no mortar which was durable when exposed to the weather, and the stone walls were therefore protected by overhanging eaves of wood, while the wooden walls needed no such shelter.

The roof shape adopted by the Dutch made dormers unpractical for light in the second story; and as metal for flashing, so essential around dormers, was scarce and difficult to obtain, dormers were usually, if not invariably, omitted, and evidently in those houses which now possess them they were added at a date far later than that of the construction of the main building. The second stories of these houses were therefore lighted at the two gable-ends only, and in several of the old buildings which remain in their original condition I have found that the second-story bedrooms were formed by partitions only, no ceilings having been constructed, so that there was a through ventilation of air from one end of the house to the other over the tops of the bedrooms. The framework was in general constructed in the same manner as in the other colonies: it was of the post and lintel type. In the earliest times the bodies of the walls were built of thick planks set edge to edge vertically; the inner sides of these planks were adzed to give a mortar clinch, and the shingles or clapboards for the exterior were nailed to the outside. The custom of filling in between the posts with studs was probably begun as early as 1725, and the spaces between the studs were often filled with brick or small stone laid up in clay; sheathing was then applied much as it is today, and the outside shingled or clapboarded, although in some instances the buildings were stuccoed directly on the studs and masonry filling between them, without sheathing or lath.

The earlier houses had little interesting detail, and, curiously enough, much of what there

BERGEN HOMESTEAD — 1655 — FLATLANDS, BROOKLYN, NEW YORK

was was strongly reminiscent of Gothic. The doorways, for example, in the old Verplanck House at Fishkill, New York, are not dissimilar from the English Elizabethan type, and hexagonal and octagonal columns were used in very many cases. The later houses, probably through the influence of the New England work, had considerable attention paid to the treatment of the doorways, the cornices and the window openings, and some of the Dutch doorways and cornices are among the most interesting Colonial works still remaining. The cornice of the main part of the Board House (which dates from 1790), for example, illustrated on pages 16 and 17, has a narrow frieze decorated in the Chinese-Chippendale manner, and the cornice of the wing shows an extremely interesting combination of dentil course and fluting; both cornices are rich, vigorous and refined. Several of the other houses have doorways carved as elaborately as could be done by a carpenter with the tools then at his command; the use of the gouge to form rosettes and other decorated forms being the marked characteristic. An excellent example of this is the doorway of the Vreeland House, which, though late in period, is much more Colonial than Neo-Grec in sentiment.

The Dutch uses of ornament were characterized, however, by the same freedom from traditions as were the masses of their houses; and indeed the pervading sentiment of all the Dutch work is one of spontaneity and disregard for precedent, rather than the adherence to formulæ customary in New England.

The Dutch houses had not, as a rule, very much pretension to stylistic correctness; they were charming rather than beautiful, and quaint rather than formal. This quality makes them especially adapted for precedents for small country houses of today, just as the symmetrical dignity of the Colonial work of New England and the South lends itself to larger and more expensive residences which may be termed mansions.

Certain of the Dutch forms, especially that of the roof, cannot be readily used, the flat slopes of the Dutch work admitting little light and air in the second story; but the other shapes of gambrel, which were used practically all over the United States, and of which there are examples existing at such widely separated points as Castine, Maine; Annapolis, Maryland; and New Orleans, Louisiana, can be harmonized with the spirit of the Dutch work with profit to our architectural design.

ROADSIDE FARMHOUSE NEAR PEARL RIVER, NEW JERSEY
Note the use of "Germantown hoods," and the fact that wings are added to the ends only.

TERHUNE HOUSE — 1670 — HACKENSACK, NEW JERSEY

The body of the house is the oldest section. One of the few examples where use was made of mouldings on the exterior other than door and window architraves.

BOARD-ZABRISKIE HOUSE, ON THE PARAMUS ROAD, NEW JERSEY

Date, 1790, carved in lintel of a cellar window. Note the Chinese-Chippendale ornament in the cornice of main house. Dormers, wing and railing probably added later.

Detail of West Wing at Right Angle to Road
BOARD-ZABRISKIE HOUSE, ON THE PARAMUS ROAD, NEW JERSEY
Of all houses in this section none is more charming; the
interest lies both in the composition and beautiful detail.

ACKERMAN (BRINCKERHOFF) HOUSE, HACKENSACK, NEW JERSEY

Date, 1704, carved in end of chimney. Interesting use of columns under the overhang in the center only.

LEFFERTS HOUSE, FLATBUSH, BROOKLYN, NEW YORK

Present house dates partly from before 1776 and partly from a century earlier. A portion of the house was destroyed by the British in the battle of Long Island, but was soon rebuilt on its undamaged beams.

JOHN PETER B. WESTERVELT HOUSE—1800—CRESSKILL, NEW JERSEY
An almost perfect example of the full development of the style.

VREELAND HOUSE AT NORDHOFF, NEW JERSEY
The wing dates from the eighteenth century; the body of the house was added about 1825, and is extremely interesting in detail, as may be seen in the frontispiece illustration.

ANDREW HARRING HOUSE AT NORTHVALE, NEW JERSEY
Rebuilt 1805 and 1838.

JAN DITMARS HOUSE—c1800—FLATLAND NECK, BROOKLYN, NEW YORK
While this house is built entirely of wood, it is interesting to note that
the proportions and type are exactly similar to the Harring House above.

VAN NUYSE-MAGAW HOMESTEAD — c1800 — FLATLANDS, BROOKLYN, NEW YORK

EARLY NINETEENTH CENTURY DUTCH HOUSE ON LONG ISLAND

Here the gambrel roof is above two full stories; unusual near New York. All existing examples thus designed have cornices and detail resembling the work of New England rather than other Dutch houses.

Doorway
LEFFERTS HOUSE ON FLATBUSH AVENUE, FLATBUSH, LONG ISLAND
Built in the seventeenth century, rebuilt about 1780. An extremely interesting
doorway, showing the freedom with which the Dutch builders used Classic motives.

Dutch Houses of
New Jersey

Text by
Clifford C. Wendehack
Photographs by
Kenneth Clark
Originally published in 1925 as White Pine Monograph
Volume XI, Number 3

KIPP HOMESTEAD—1800—ON THE BANK OF THE HACKENSACK RIVER

EARLY DUTCH HOUSES
OF NORTHERN NEW JERSEY

THE early architecture found in New Jersey, more particularly on the banks of the Hackensack River, stands apart and distinct from all other types of early American domestic architecture in the United States.

These houses with their quaint gambrel roofs, wide overhanging eaves and broad flat walls of brown stone, have usually been designated as Dutch Colonial although this term is incorrect and misleading.

The best examples of the style peculiar to northern New Jersey were erected long after 1664 when Manhattan Island and the surrounding country had ceased to be a Dutch colony. The style developed slowly, reached maturity in the beginning of the eighteenth century, and attained its highest perfection after the War of Independence.

There is a type of true Dutch Colonial architecture, found in South Africa in the Dutch and Boer settlements which was developed when the veldt was still under Dutch rule. It is a very simple style with low broad walls and simple openings, but it never attained the charm and proportion of the style developed in America.

The Dutch descendants who settled in northern New Jersey, had very little intercourse with their English neighbors to the south in the region of Philadelphia. Living in an isolated manner, these hardy northern settlers expressed in their homes their tenacious individuality, and for several generations adhered very closely to a set type of plan and design. Whether their houses were large or small the same characteristics can be seen in all the work done in this section.

The history of this particular portion of New Jersey along the banks of the Hackensack River, from the town of Hackensack to the New York state line, is most interesting particularly that bearing on the houses under consideration.

The Delaware Indians, among whom was a tribe called the Hackensacks, caused the early settlers considerable trouble, and it was not until the last quarter of the seventeenth century, about 1670, that John Berry an Englishman, and a French Huguenot named De-

marest, were each granted an extensive tract of land, comprising what is now the town of Hackensack and the surrounding country. This tract was soon occupied by Dutch settlers from Manhattan. Records bear the date of 1686 when these settlers began to unite into community life, and the beginning of the next century found many prosperous communities of busy farmers with the Dutch habits of thought and expression.

With this element, mingled the French Huguenot strain, descendants from the Demarests, which, however, never became strong enough to dominate the Dutch influence.

Slavery was introduced in New Jersey in 1664, and the land owners were encouraged to increase the number of slaves by a bonus of seventy-five acres of additional land for each slave brought into the colony. Records show that in Bergen County alone there were twenty-three hundred slaves in 1790. This fact has a very important bearing on the architecture of this period, for, being possessed of unlimited, inexpensive labor, the Dutch tendency towards solidity could be indulged in to the utmost. The stone used for the walls of their houses, was in all probability cut and laid by slave labor.

The stone employed is usually called brownstone, and is to be found in great profusion throughout New Jersey. In many localities this brown sandstone is found in ledges and was easily adaptable to the solid flat beds so characteristic of this early masonry.

Stone of irregular size laid in random bond was used generally until the beginning of the nineteenth century when it became a practice to employ cut stone for at least the front and corners with uniform jointing which produces a modern effect in some of the later examples.

The Demarest House on the Saddle River (pages 40 and 41) is probably one of the most effective illustrations of highly developed stone work. On the rear and ends is demonstrated the earliest type of bonding and irregular laying of stone, and by observation it will be seen that the horizontal beds are maintained throughout in a most exacting manner.

There is abundant proof that practically all of this early masonry was originally laid in ordinary clay taken from the surrounding fields and mixed with straw. The quality of the New Jersey soil was particularly adapted to this use. Stone work laid in red clay mortar was practically impervious to moisture, and many examples of clay joints in window sills are still intact. At later periods of restoration, most of this stone work was pointed up with lime mortar producing the white joints visible in the accompanying illustrations.

The most characteristic feature of these houses is the enormous overhang and sweep of the eaves. Undoubtedly the Dutch thrift of maintaining property without repairs was responsible for the development of this very charming construction, as it was the means by which water was diverted from the walls of the buildings and prevented the washing out of the clay joints in the masonry.

Assuming that this theory of the overhanging eaves is correct, it is not easily explained why the gable walls were made so extremely flat, in fact, more so than in any other type of early American architecture. However, there was some thought taken to the condition of the gable walls by the fact, hardly without exception, that the gables were covered with shingles and later with clapboards. Thus for purely utilitarian reasons there was developed one of the most charming architectural compositions.

The overhanging roofs were sheathed on the under side with boards and were usually carried out to form a boxed cornice five or six inches deep on the outer edge. On the gable ends a delicately moulded skirt border was generally used and occasionally a small row of dentils is to be found.

When the original cottage grew too small for its owner, a second larger house was built against one end, and in most cases became the main portion of the house and the original building was often changed into a kitchen wing. Frequently a second wing was added to balance the original house. Thus the form of plan so often found was the result of natural growth of the size and prosperity of the family.

A reversed order of building is found in the Terhune House in Anderson Street, Hackensack, New Jersey. The main portion of this house, which is built of brown stone and now whitewashed, was built by John Terhune in 1670. In 1800 a descendant of the original owner occupied the house and added a frame wing for kitchen and dining room use.

Irrespective of the period in which the earliest portions of these houses were constructed, there seems to have been a faithful adherence to tradition as to lines and roof pitches which has created an exquisite mass so characteristic of the region. The steeper slope is slightly less than forty-five degrees and the top slope ranges from twenty to twenty-three degrees and is usually quite short.

Around a tradition which permitted in the beginning only the use of local standstone, many interesting variations in architectural forms employing the native pine are to be found. One of the most interesting is the employment of square wood columns. The original excessive overhang was transformed into a porch and undoubtedly many of the examples, where these narrow columned porches are to be found, were added to at a later date.

The Kipp Homestead on the banks of the Hackensack River is one of the most interesting examples left to us both historically and architecturally. In the gabled end facing the river, we find an example of the natural tendency of these builders to employ masonry wherever possible. We are told that the bricks in this wall were brought from Holland by an early Dutch trader as ballast. The opposite gable is of clapboards; the front and rear walls are surmounted by a most unusual cornice of delicately carved modillions.

Toward the beginning of the nineteenth century, the treatment of doorways, cornices and window openings were given added consideration as may be seen in the Vreeland House (pages 31–38). It is interesting to note in this example the clever use of one and the same set of mouldings throughout and to see what rich and vigorous carving was done with a gouge.

The Lincoln House in Hackensack, originally built by an uncle of John Terhune in 1773, is an excellent demonstration of the softening effect of round pine columns and the more abundant use of wood cornices and gables made possible by a modern restoration of the original sand stone house. The cornice, details and roof lines have been faithfully carried out from the original.

To appreciate these old homes of New Jersey at their best, one must view them in their native landscape amid the rolling slopes and green meadows which have formed their setting for so many generations.

The mellowing effect of time and the elements has brought out the texture of the stone work and in many cases the original wood clapboards or shingles have been left to us without paint, and have mellowed to a golden russet as may be seen on the gable end of the Ackerman House on Polifly Road, Hackensack. (See page 27.)

It would be a noble work indeed if the New Jersey State Board of Architects could acquire some of these finest examples and preserve them intact without mutilation for posterity.

ACKERMAN HOUSE, POLIFLY ROAD, HACKENSACK, NEW JERSEY

HENDRICK BRINCKERHOFF HOUSE, TEANECK, NEW JERSEY

TERHUNE HOUSE—1670—ANDERSON AVENUE, HACKENSACK, NEW JERSEY
Built by John Terhune.

NUMBER 220 GRAND AVENUE—1803—ENGLEWOOD, NEW JERSEY

The oldest house in Englewood.

VREELAND HOUSE—1818—NORDHOFF, NEW JERSEY

Front Doorway
VREELAND HOUSE, NORDHOFF, NEW JERSEY

Detail of Front Elevation
VREELAND HOUSE, NORDHOFF, NEW JERSEY

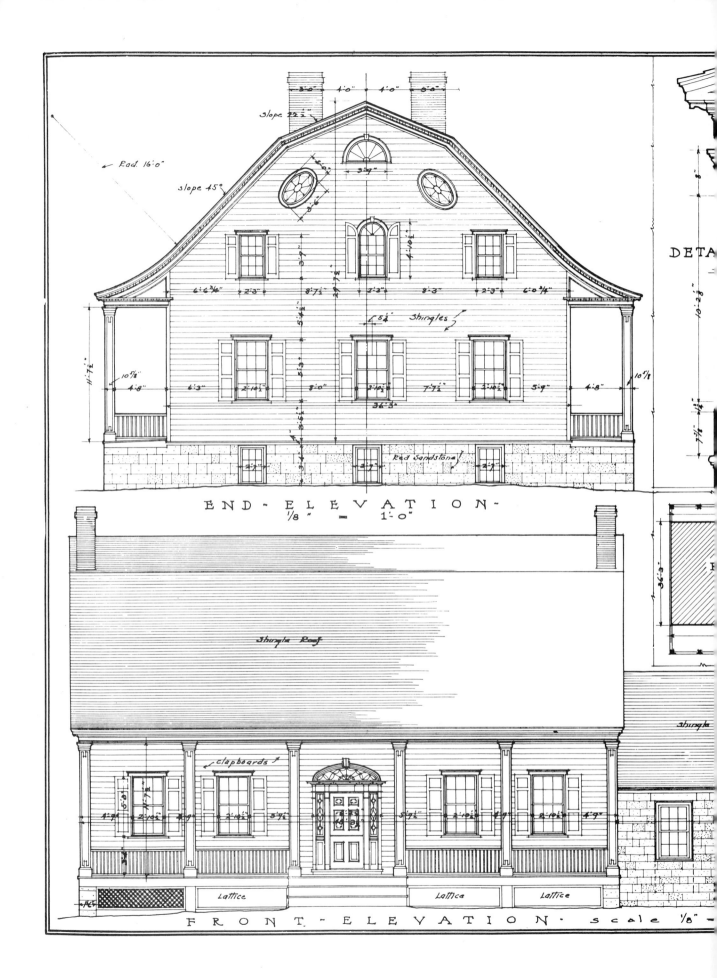

· E N D · E L E V A T I O N ·
1/8" = 1'-0"

· F R O N T · E L E V A T I O N · scale 1/8" =

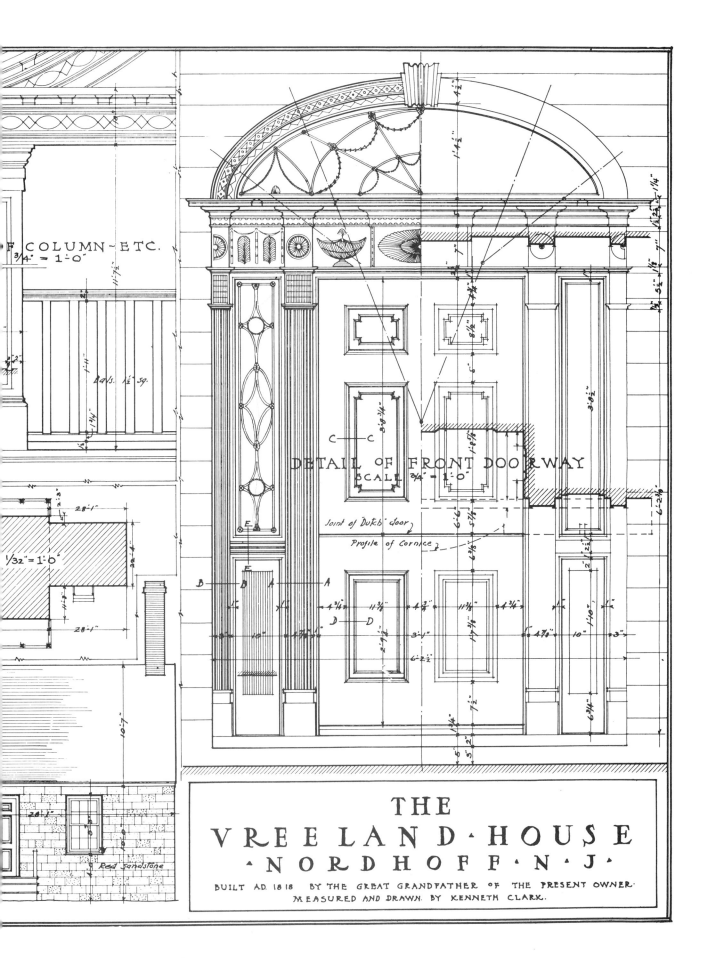

F COLUMN ~ ETC.
3/4" = 1'-0"

DETAIL OF FRONT DOORWAY
SCALE 3/4" = 1'-0"

Joint of Dutch door

Profile of Cornice

Rails 1½ sq.

1/32" = 1'-0"

Red Sandstone

THE
VREELAND·HOUSE
·NORDHOFF·N·J·

BUILT A.D. 1818 BY THE GREAT GRANDFATHER OF THE PRESENT OWNER.
MEASURED AND DRAWN BY KENNETH CLARK.

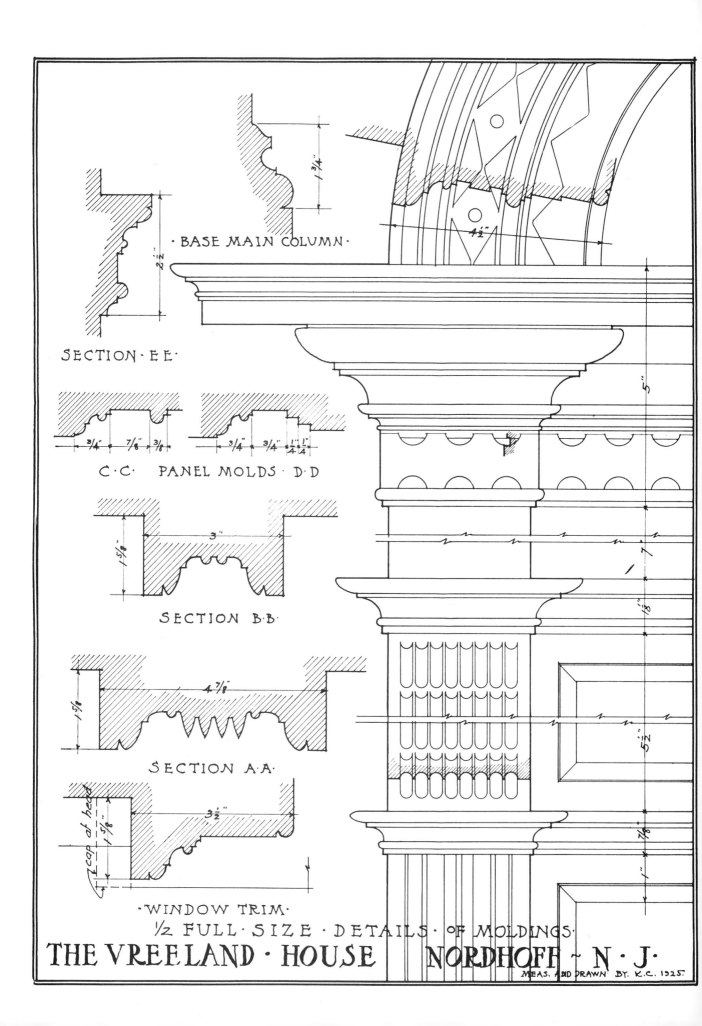

· BASE MAIN COLUMN ·

SECTION · E·E ·

C · C   PANEL MOLDS · D·D

SECTION · B·B

SECTION · A·A

· WINDOW TRIM ·

½ FULL · SIZE · DETAILS · OF MOLDINGS ·

THE VREELAND · HOUSE   NORDHOFF ~ N · J ·

MEAS. AND DRAWN BY. K.C. 1925.

Detail of Front Doorway
VREELAND HOUSE, NORDHOFF, NEW JERSEY

End Elevation
VREELAND HOUSE—1818—NORDHOFF, NEW JERSEY
Built by the great-grandfather of the present owner.

HOPPER HOUSE — 1816–1818 — POLIFLY ROAD, HACKENSACK, NEW JERSEY

Rear Elevation

DEMAREST HOUSE—1837—ON THE SADDLE RIVER, NEW JERSEY

End Elevation

DEMAREST HOUSE — 1837 — ON THE SADDLE RIVER, NEW JERSEY

JOSEPH C. LINCOLN HOUSE, HACKENSACK, NEW JERSEY
Restoration of the original house built by an uncle of John Terhune in 1773

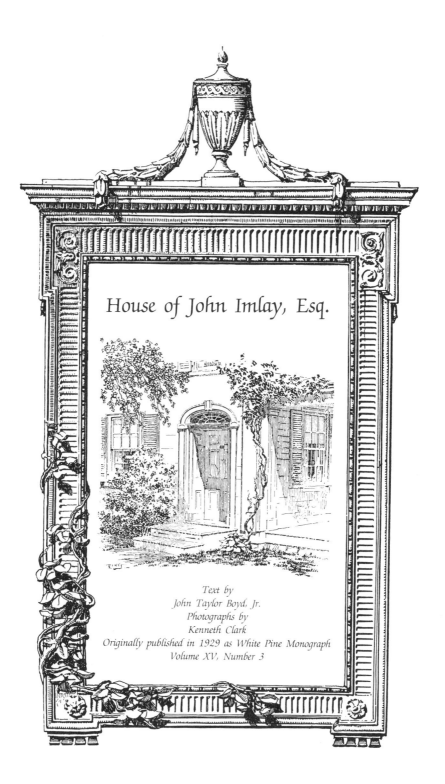

House of John Imlay, Esq.

Text by
John Taylor Boyd, Jr.
Photographs by
Kenneth Clark
Originally published in 1929 as White Pine Monograph
Volume XV, Number 3

JOHN IMLAY HOUSE, ALLENTOWN, NEW JERSEY

THE HOUSE OF JOHN IMLAY, ESQ.
ALLENTOWN, NEW JERSEY

THE rare quality of the Imlay House, and particularly the extraordinary interest of its interior, deserve to make it widely known. Among the many early American houses remaining in New Jersey, this house has scarcely a rival and it would take high rank among old houses outside the state.

Connoisseurs of early Americana know of the Imlay House as the source of the wall paper now in one of the rooms of the American Wing of the Metropolitan Museum of Art, New York. Architects who are familiar with our native tradition doubtless have noted the brief mention of the Imlay House made in the excellent book, *Colonial Architecture for those about to Build*, by Wise and Beidleman. But such mention is entirely too scant; only those few who have actually seen the house itself have any idea of its distinction.

Nor is there much knowledge available concerning the Imlay family. John Imlay (b. 1754–d. 1813), who built the house, was a shipping merchant engaged in the West Indies trade, with offices in Philadelphia, but who retired to the placid countryside of New Jersey for his permanent home. The writer has a slight knowledge of the Imlays, since his grandmother was an Imlay, whose father, Robert Imlay, was likewise a prosperous Philadelphia merchant, closely related to the Imlays of Allentown. Another Imlay was Gilbert, captain in the Continental Army, who, after the Revolution, went to Paris and there entered into a love affair with Mary Wollstonecraft, the English writer and pioneer feminist, whose daughter married the poet Shelley.

The date of construction of the house must fall close after 1790, because the bill of sale for the wall paper now at the Metropolitan Museum, bears the date of April 18th, 1794. The paper was sold by William Poynsett of Philadelphia, a well-known maker and importer of wall papers. It is an imported French Louis XVI, hand blocked, classic design, in a pattern of light grey garlands and small medallions displayed against a reddish-brown background, delicate in scale.

In those days of the first decade of the Republic, Allentown consisted of about three substantial houses built in a group close to the elm-lined road, and in addition a gristmill that was situated about two hundred yards away. Allentown lies ten miles east of Trenton and about forty miles northeast of Philadelphia. The Trenton district was important at that time and Jerome Bonaparte made his first residence in America at Bordentown, six miles away from Allentown. Today, Allentown has changed but little from the idyllic days following the Revolution. It has grown to about a score of houses and the old gristmill still runs! Requiescat!

The Imlay House is placed lengthwise of the road and is brought so close to the street line that the lowest of the four marble entrance steps encroaches on the sidewalk. The two short, balanced wings are set back, permitting a small lawn closed off from the street by a picket fence. The corners of the house are placed to the four cardinal points of the compass. The plan of the main house is the familiar one of an entrance opening from the street into a central stair hall running entirely through the house, with four main rooms, two on each side. These main rooms, about twenty feet square, are formed by a transverse wall and the two big chimneys are placed back to back, to serve the front and rear rooms.

One of the wings is the kitchen and the other was John Imlay's office. Each has a chimney in the gable end. The house is abundantly supplied with open fireplaces, there being eleven all told, each one furnished with a mantel, and each mantel different in design from the others. The more important rooms are variously decorated with low wainscot, fine door and window trim and ornamental cornice at the ceiling.

The house is solidly constructed, having stone foundations that are carried nearly up to grade, topped with several courses of brick, forming a low base which is finished with a moulded brick watertable. The interior division walls of the cellar are of brick.

The main house and the wings are roofed with ridges running the long way of the house. The main roof is pierced with three small gabled dormers, placed close to the eaves and designed with arched windows and small pilasters. Except for a small porch which has been added to the kitchen wing, the house remains today practically unchanged.

Such are the main characteristics. It is when one examines the carefully wrought, classic detail that one sees what a masterpiece it is, this serene old home of the Early Republic. Conventional as the exterior is, one could never tire of its simple, frank proportions. It is trite to say that every line, detail and moulding is right and could not be changed without damage to the effect, but it is true in this case. The broad low proportions, the perfect scale of the two rows of windows, the delicately detailed cornice and above all the beautiful doorway, with its jewel-like sparkle of decoration—these can hardly be overpraised.

The architectural elements of the Imlay House, are fairly typical of the period and of the region northeast of Philadelphia. Every bit of detail, down to the tiniest curve and moulding has been studied almost to the ultimate possibility of craftsmanship. As a result of this exceptional design and workmanship, the Imlay House is like one of those finest masterpieces of early American furniture—say one of Duncan Phyffe—which has a final touch of grace, of perfection beyond other similar pieces.

The shutters are thicker and have more character in the panel mouldings than is usually the case. On the main doorway, the delicate play of light and shade on its projecting forms reveal much subtlety that is not at first apparent. Note in the illustration the tiny drill-holes accenting the vertical channelings and the under sides and front of the dental course under the raking cornice, also the wood pattern of the fanlight with the three little "drumsticks" in the center, and the moulded edge of the door jamb, which is stopped by a tiny block just below the impost of the arch.

The main cornice shows the same imagination and the same infinite care. The edges of the clapboards and of the gable are beaded, in order to soften the lines; the vertical corner-board is paneled and is more strongly modeled than usual; and the graceful brackets under the overhanging cornice have an unusual detail, particularly at the corner. The details of the dormers are slightly more accented, in order, no doubt, not to lose their character when seen from the street. Another variation is seen in the change in size of the window panes on the ends of the main portion of the house and in the wings. On the front, it will be noticed that the siding is matched in order to present a more finished surface close to the street. This siding has a ¼" bead and is about 5⅝" wide. The end clapboards are laid about 8" to the weather.

But, as I have intimated, remarkable as the exterior of the Imlay House is, even more interesting is its interior.

The interiors have all the sophistication and fine finish of the best early Republican houses of the mansion type, as found in the north. They have not the low-ceil-inged, homely air of the farmhouse type on the one hand, nor have they the Georgian massiveness of the great Southern plantation homes. They seem to stand midway between these extremes. Indeed, John Imlay's house bears much resemblance to the homes of his contemporary merchant ship masters of Salem and Newburyport and, in detail, one senses the same classic hand in the work of the craftsmen or architects who built this house—probably men from Philadelphia—as one feels in the Salem interiors of Samuel McIntire.

In all, the house contains fifteen rooms, of which the most important naturally are those on the ground floor of the main portion of the house. It is a curious fact that the interior cornices have almost the same design as the cornice of the exterior, although their scale, of course, is smaller. Points of interest are the original tile fireplace facings in the two main front rooms and the hardware, of which the locks are generally the box-lock type, of wrought iron with brass handles.

The detail of the main stair hall has a touch of farmhouse architecture, though it is quite perfect in the design of rail, balusters and stairs. An unusual form is the graceful scroll-saw ornament under each riser.

One could hardly imagine a finer interpretation of the style than is seen in the decoration of the drawing room. This work is indeed comparable to McIntire's, and it was here that just a little more effort was spent, if possible, than in the others. Its rare beauty is attained by scholarly design, by the rich details and by the same painstaking workmanship that is noted throughout. This is apparent in the illustration of the detail of the upper part of the over-mantel, showing the room cornice. The parlor is less sophisticated, particularly with regard to the cornice, than the drawing room, but it is no less perfect and has a charm of its own. John Imlay's portrait hangs over the mantel.

Upstairs, one finds a different treatment in the guest room of the second story front. The cornice is heavier and the detail has more relief. The mantel is almost square in shape to fit the smaller fireplace of a bedroom and it does not extend across the full width of the chimney-breast, but leaves a space for a row of long panels on each side. The original wall paper is still on the walls. It is of a freer design than the one in the Metropolitan Museum, but equally delicate and finished with dark border at the top under the wood cornice.

There is an endless fascination about an old home like this. Its quiet beauty is typical of the early Republican days of serenity and good living. Our age is different but perhaps this very fact makes us appreciate all the more a fine example of the admirable architecture of earlier times, such as the Imlay House.

JOHN IMLAY HOUSE—1790—ALLENTOWN, NEW JERSEY

Office Wing
JOHN IMLAY HOUSE, ALLENTOWN, NEW JERSEY

Doorway
JOHN IMLAY HOUSE, ALLENTOWN, NEW JERSEY

Detail of Entrance Doorway
JOHN IMLAY HOUSE, ALLENTOWN, NEW JERSEY

Detail of Cornice
JOHN IMLAY HOUSE, ALLENTOWN, NEW JERSEY

Detail of Dormer Window
JOHN IMLAY HOUSE, ALLENTOWN, NEW JERSEY

Drawing Room Mantel
JOHN IMLAY HOUSE, ALLENTOWN, NEW JERSEY

Detail of Drawing Room Cornice and Over-Mantel
JOHN IMLAY HOUSE, ALLENTOWN, NEW JERSEY

Detail of Drawing Room Mantel Shelf
JOHN IMLAY HOUSE, ALLENTOWN, NEW JERSEY

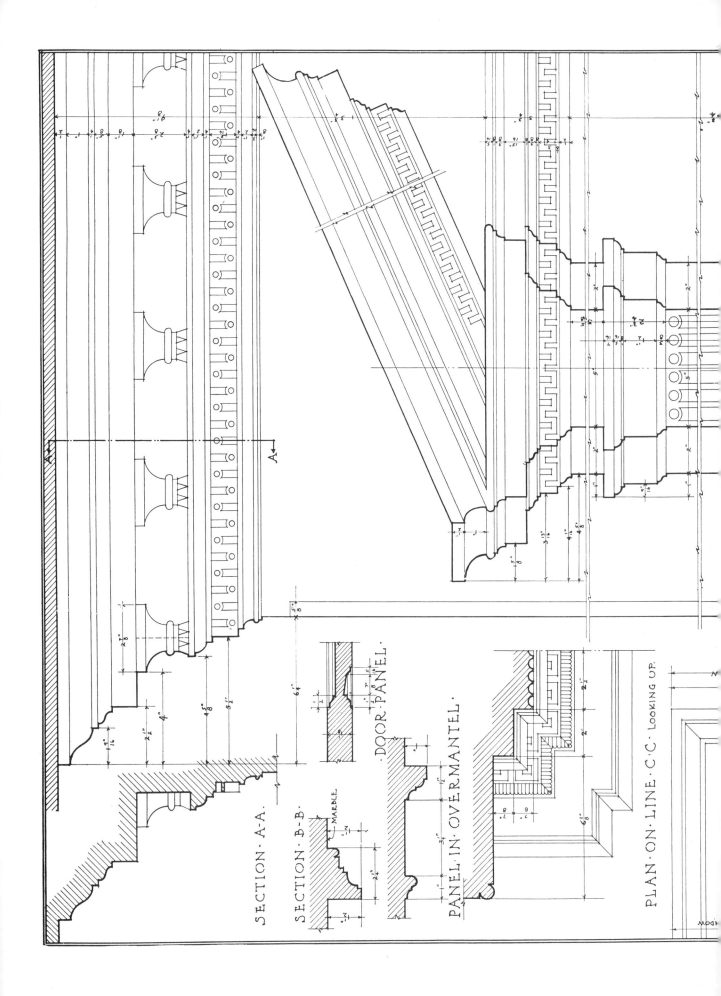

SECTION · A-A ·

SECTION · B-B ·

· DOOR · PANEL ·

· PANEL · IN · OVERMANTEL ·

· PLAN · ON · LINE · C·C · LOOKING UP ·

MARBLE

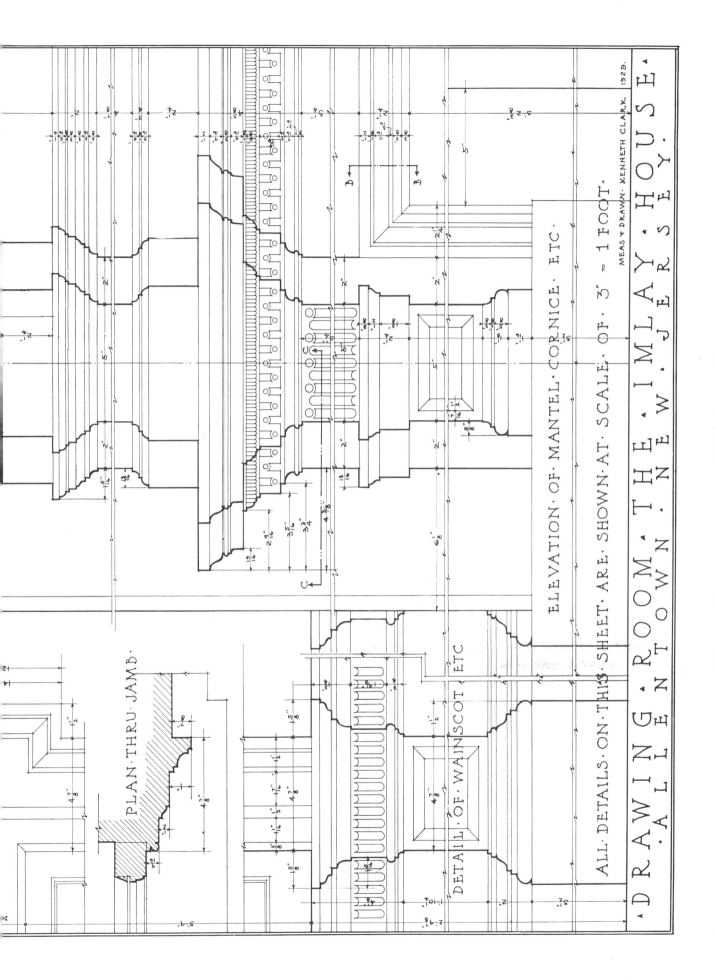

PLAN·THRU·JAMB·

ELEVATION·OF·MANTEL·CORNICE·ETC·

DETAIL·OF·WAINSCOT·ETC

ALL·DETAILS·ON·THIS·SHEET·ARE·SHOWN·AT·SCALE·OF·3"·=·1·FOOT·

MEAS·+·DRAWN·KENNETH·CLARK· 1929.

·DRAWING·ROOM·THE·IMLAY·HOUSE·
·ALLENTOWN·NEW·JERSEY·

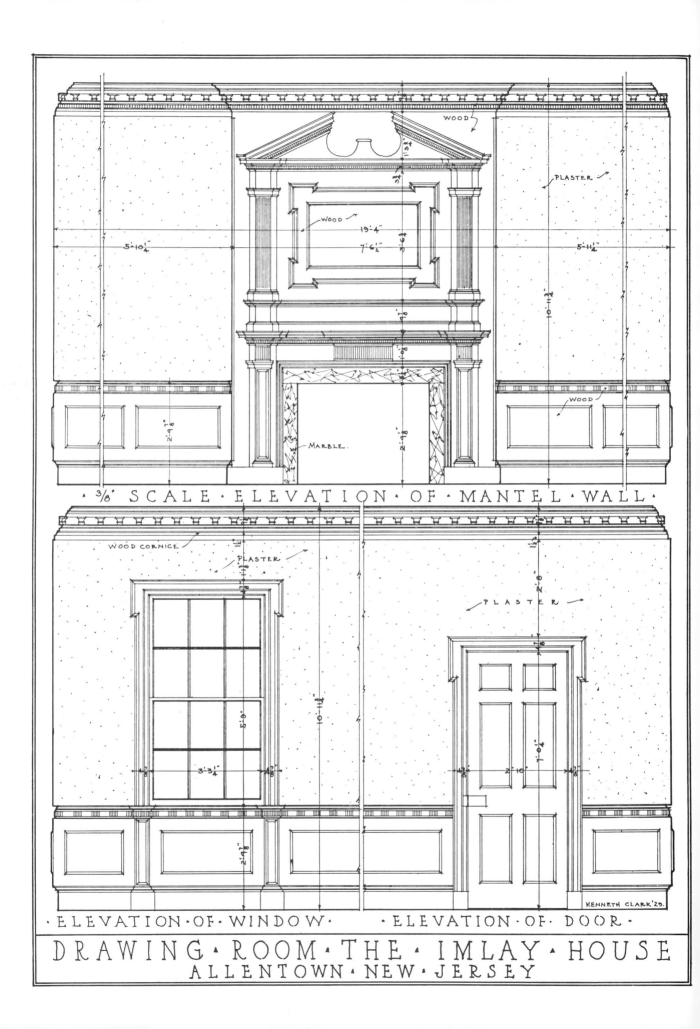

WOOD

PLASTER

WOOD

5'-10¼"

19'-4"

7'-6½"

3'-6¼"

5'-11¼"

10'-11¾"

2'-9⅞"

2'-9⅞"

MARBLE.

WOOD

· ⅜" · SCALE · ELEVATION · OF · MANTEL · WALL ·

WOOD CORNICE

PLASTER

PLASTER

10'-11¾"

5'-9"

3'-3¼"

4⅞"

4⅞"

7'-0¼"

4⅞"

2'-10"

4⅞"

2'-9⅞"

KENNETH CLARK '29.

· ELEVATION · OF · WINDOW ·   · ELEVATION · OF · DOOR ·

DRAWING · ROOM · THE · IMLAY · HOUSE
ALLENTOWN · NEW · JERSEY

Parlor Mantel
JOHN IMLAY HOUSE, ALLENTOWN, NEW JERSEY

Detail of Stair Well
JOHN IMLAY HOUSE, ALLENTOWN, NEW JERSEY

Detail of Newel and Stairs
JOHN IMLAY HOUSE, ALLENTOWN, NEW JERSEY

Lower Hall Doorway
JOHN IMLAY HOUSE, ALLENTOWN, NEW JERSEY

Guest Bedroom Mantel
JOHN IMLAY HOUSE, ALLENTOWN, NEW JERSEY

Detail of Guest Bedroom Mantel
JOHN IMLAY HOUSE, ALLENTOWN, NEW JERSEY

Detail of Upper Stories
JOHN IMLAY HOUSE, ALLENTOWN, NEW JERSEY

Detail of Window and Shutters
JOHN IMLAY HOUSE, ALLENTOWN, NEW JERSEY

Burlington County
Court House

Text by
Fenimore C. Woolman
Photographs by
Kenneth Clark
Originally published in 1926 as White Pine Monograph
Volume XII, Number 3

BURLINGTON COUNTY COURT HOUSE, MOUNT HOLLY, NEW JERSEY

THE BURLINGTON COUNTY COURT HOUSE
AT MOUNT HOLLY, NEW JERSEY

INSPIRING history often takes place in buildings of inspiring architecture—or inspiring architecture creates an atmosphere condusive to high and ennobling events, as the reader likes. In either case, probably the most famous early American public buildings are the group surrounding Independence Square in Philadelphia, a group composed, you will remember, of the dignified State House, with its arcades and wings, designed by the great amateur architect, Andrew Hamilton, now known as Independence Hall, and the two buildings that complete the group—the Philadelphia County Court House and the City Hall, erected in 1787. Congress sat in the latter building from 1790 to 1800.

In those early days some knowledge of architecture was considered an essential part of every gentleman's education, for at least two reasons; to round out his general knowledge and as a practical tool when the time to build his own dwelling or to assist in the planning of some public building came. The trained professional architect of today was almost unknown. It is not surprising therefore to find that Major Richard Cox, Zachariah Rossell and Joseph Budd, having been entrusted in 1796 with the building of a new county Court House for Burlington County, New Jersey, should have been able to emulate the creator of the Philadelphia City Hall, its dignity and repose having aroused their admiration, and that they or their appointed craftsmen were successful in catching the spirit of the earlier building and adapting it to their own needs and purposes. The Court House which these amateur New Jersey architects conceived is not as well known to the architectural student of today as is Congress Hall—not because they were unsuccessful in their purpose to create a beautiful and well constructed building, but rather because Mount Holly is in a "sand hole" in West Jersey, if we use the geographical term used in the early days, several miles from the old city of Burlington, the first capital of West Jersey, and not on the beaten path of the architectural explorer.

The Court House at Mount Holly, is nevertheless, one of our priceless American architectural inheritances, standing virtually as it was built, an enduring memorial to the most elegant period in early American architectural history. Its very dignity precludes any idea of unseemly conduct or lack of majesty in the administration of the business of meting out justice—yet it was built only after violent and bitter contention.

According to the records, the original county Court House, which was built in the city of Burlington on the shores of the Delaware River in 1690, was in such a state of dilapidation by 1795, that a large expenditure of money was required to repair it, which money the freeholders refused to vote. An act of legislature was passed, enabling the people to erect a new court house, and authorizing a vote of the people of Burlington County to determine where it should be built, as many thought the public buildings should be nearer the center of population than Burlington, which is at one side of the county. The places nominated were Black Horse, Mount Holly and Burlington. Mount Holly won after much excitement and a hotly contested election. Dr. Read, in his "Annals of Mount Holly" says: "I am informed that at the election held at the Mount Holly Market House, to say where the court house should be built, that some voters came from the pine districts with dingy faces, and after voting, retired to the saw mill race, washed themselves, returned and voted again. This may have been true, but for want of proper authentication, we may consider it the product of a romantic imagination and place no reliance upon it." But, such propaganda, evidently rife at the time, had its effect in keeping alive feelings of resentment, however amusing it may be a century and a half later.

The citizens of Burlington were so aggravated by the loss of the court house that it would have been unwise, if not actually unsafe for a Mount Hollian to show himself in Burlington. The indignity of being shorn of their pride was never forgiven or forgotten by the old citizens of Burlington, but now time has healed the wound.

The Mount Hollians being victorious, the court

house was erected in 1796 on land purchased in that year from Joseph Powell for 210 pounds. The contracting carpenter for the building was Michael Rush, a native of Mount Holly. Samuel Lewis was a master carpenter and assisted in and superintended the construction. Undoubtedly they were aided in their work by a careful study of the City Hall in Philadelphia and by the early architectural reference books containing measured drawings of Georgian details which appeared in America toward the middle of the eighteenth century.

Like the little building in Philadelphia, the court house is an oblong structure of brick with marble and white wood trim, two stories high, hip roofed and surmounted in the center by a well proportioned octagonal open cupola. On the front a pediment springs from the cornice over a slightly projecting central section of the façade. Unlike the Congress Hall, in that, instead of a three sided bay breaking the rear wall, it has a semicircular wall to form the end of the court room.

We feel that the "designers" of the Mount Holly building have been more successful in the treatment of the stoop and doorway feature than their brethren in Philadelphia. The simple dignity and scale of the doorway with its graceful fanlight above are in accord with the round headed windows of the lower story. These windows are set effectively in the brick arched openings with marble sills and keystones. Like Congress Hall, the round topped windows have sliding sash with twelve-paned lower sashes and upper sashes with ten small ornamental panes to make up the semicircle above twelve rectangular panes. The upper story windows have twenty-four panes except the one over the entrance, which has thirty panes. They are square headed with flat brick arches and marble keystones.

The brick walls are surmounted by a hand tooled wood cornice, its coved member having a series of recessed arches and the Grecian band or double denticulated moulding beneath. At the second floor level a white

marble belt course has been used. The coat-of-arms cut in marble and set in the wall over the front door was a later addition, being the gift of one Isaac Hazalhurt.

In 1807 the Surrogate's office to the north and the Clerk's office to the south of the court house were built, making a most attractive group.

No less interesting than the outward appearance is the aspect of the spacious hall with its beautiful staircase with a half-way landing. As originally planned there was a large folding door in the center of the hall, opening into the court room; on each side of the door there was a gallery with seats one above the other for spectators. In the center of the court was a platform elevated some twelve inches above the aisle and surrounded by a high railing with turned balusters of slender grace. The space was used for the grand and petit jurors. The illustration shows the court room as it is today with a rearrangement of the spectators' seats, railing, desks, jury box, etc.

The climb into the cupola from the upper story disclosed the generous roof timbers obtained in the days when Mount Holly was on the edge of the great forest area known as the Pines or Pine Barrens. The old name Pine Barrens implies something like a desert but as a matter of fact, the region produced magnificent forest trees. The original growth, pine in many places, consisted also of lofty oak, hickory, gum, ash, chestnut, etc., interspersed with dogwood, sassafras and holly and in the swamps the valuable cedar.

Like the early houses of West Jersey, the Court House at Mount Holly is of simple, well proportioned architecture of a distinctive type, less luxuriant, massive and exuberant than that across the river in Pennsylvania although both were evidently derived from the Christopher Wren school. The old court house seen today in its setting of ancient and still flourishing button-wood trees, seems to reflect faithfully the simple feeling of the Quaker people.

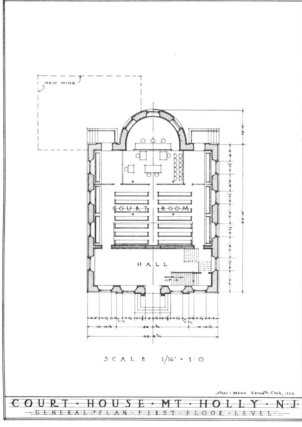

SCALE 1/16" = 1.0

MEAS. DRAWN Kenneth Clark, 1926

COURT · HOUSE · MT · HOLLY · N·J·
GENERAL · PLAN · FIRST · FLOOR · LEVEL

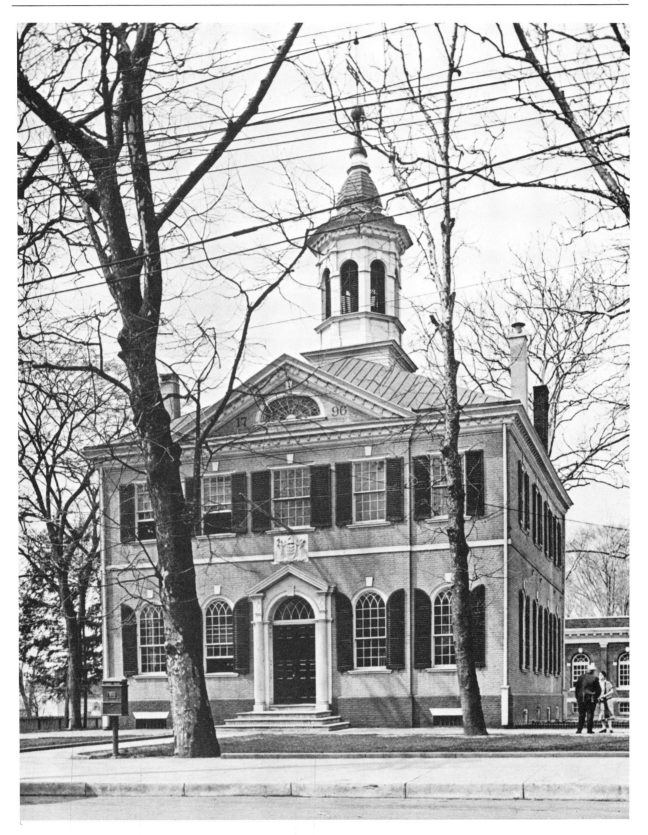

BURLINGTON COUNTY COURT HOUSE, MOUNT HOLLY, NEW JERSEY

BURLINGTON COUNTY BUILDINGS, MOUNT HOLLY, NEW JERSEY

Interior of Court Room

BURLINGTON COUNTY COURT HOUSE, MOUNT HOLLY, NEW JERSEY

Detail of Stairway
BURLINGTON COUNTY COURT HOUSE, MOUNT HOLLY, NEW JERSEY

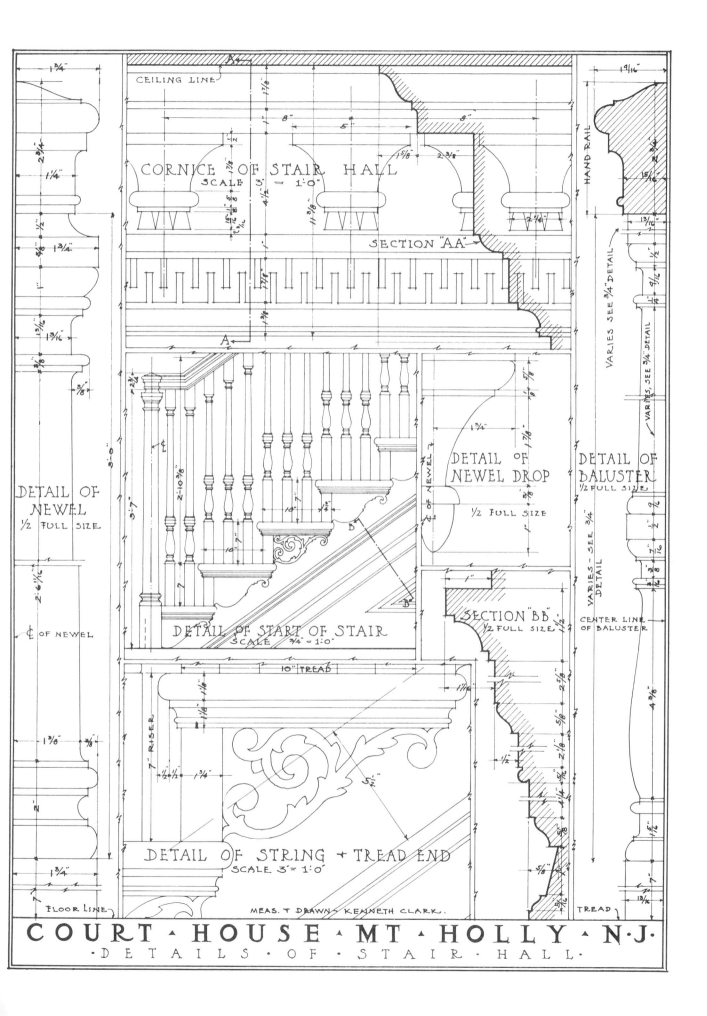

CEILING LINE

CORNICE OF STAIR HALL
SCALE 3" = 1'0"

SECTION "AA"

A

HAND RAIL

VARIES SEE 3/4" DETAIL

VARIES, SEE 3/4 DETAIL

DETAIL OF
NEWEL
1/2 FULL SIZE

DETAIL OF START OF STAIR
SCALE 3/4" = 1'0"

DETAIL OF
NEWEL DROP
1/2 FULL SIZE

DETAIL OF
BALUSTER
1/2 FULL SIZE

VARIES – SEE 3/4 DETAIL

SECTION "BB"
1/2 FULL SIZE

CENTER LINE
OF BALUSTER

₵ OF NEWEL

10" TREAD

7" RISER

DETAIL OF STRING + TREAD END
SCALE 3" = 1'0"

FLOOR LINE

MEAS. T DRAWN KENNETH CLARK.

TREAD

COURT · HOUSE · MT · HOLLY · N·J·
· D E T A I L S · O F · S T A I R · H A L L ·

Rear Entrance and Wall of Court Room
BURLINGTON COUNTY COURT HOUSE, MOUNT HOLLY, NEW JERSEY

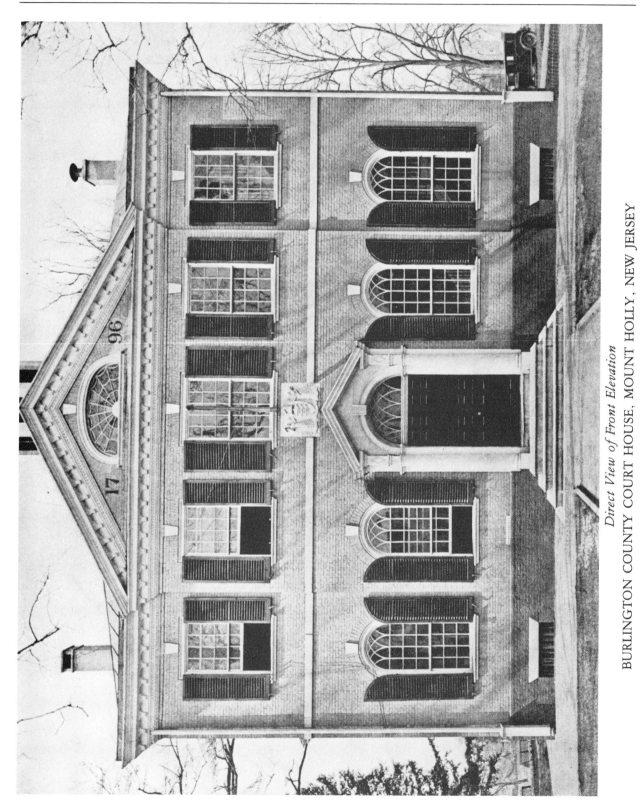

Direct View of Front Elevation
BURLINGTON COUNTY COURT HOUSE, MOUNT HOLLY, NEW JERSEY

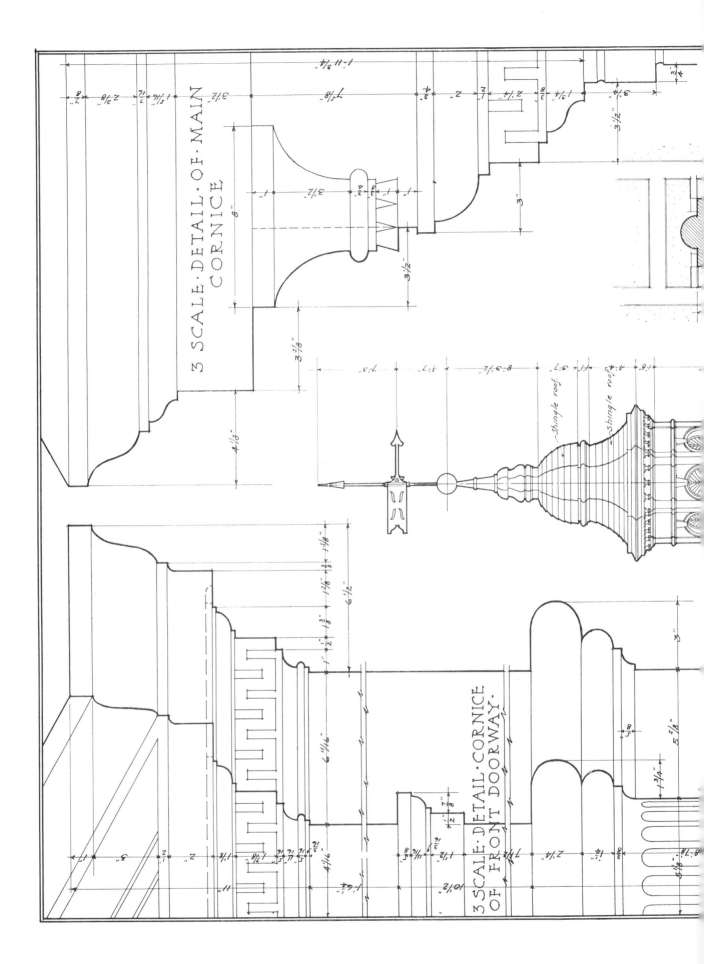

3 SCALE · DETAIL · OF · MAIN CORNICE

3 SCALE · DETAIL · CORNICE OF · FRONT DOORWAY.

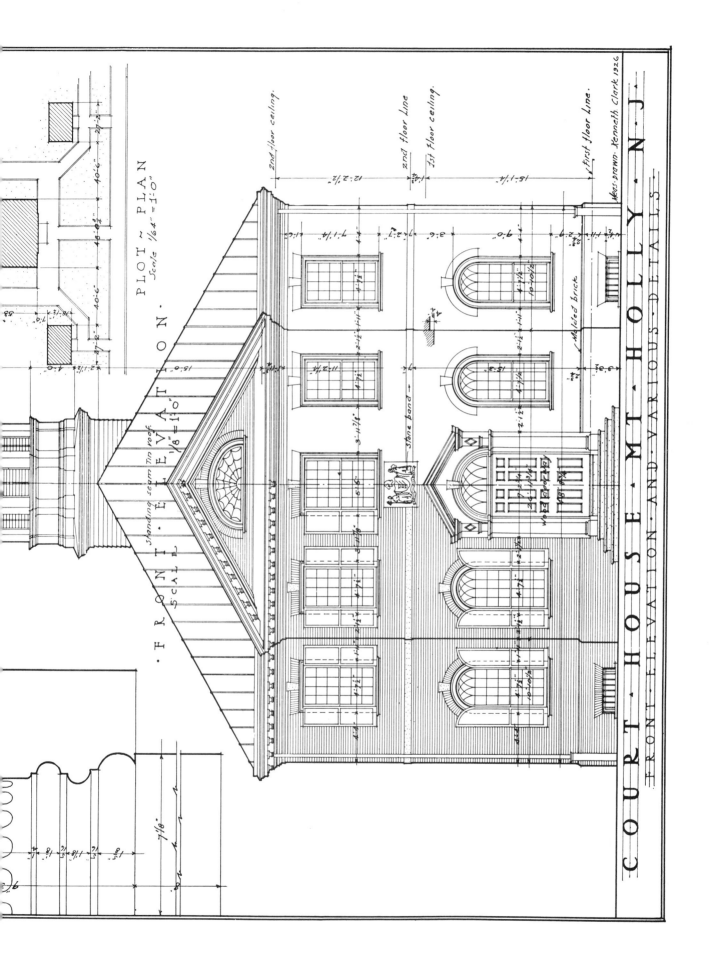

PLOT ~ PLAN
Scale 1/64" = 1'-0"

FRONT ELEVATION.
Scale 1/8" = 1'-0"

Standing seam Tin roof.

2nd floor ceiling.
2nd floor Line.
1st floor ceiling.
First floor Line.

Milled brick.
Stone band →

Meas. drawn. Kenneth Clark 1926

COURT · HOUSE · MT · HOLLY · N · J ·
FRONT · ELEVATION · AND · VARIOUS · DETAILS

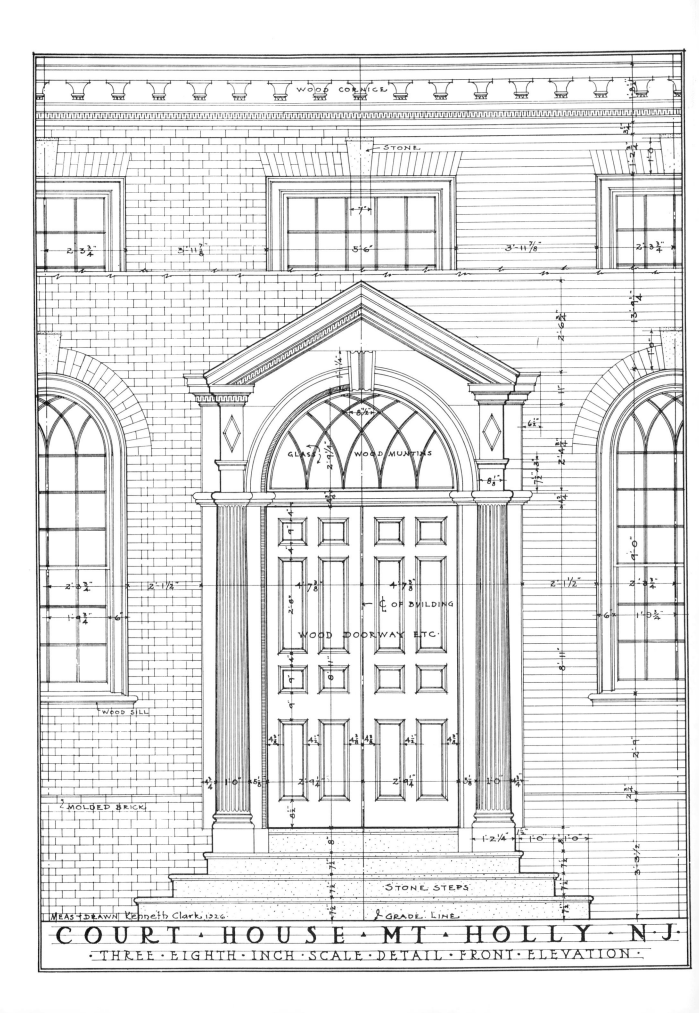

WOOD CORNICE

STONE

5'-6"

GLASS

WOOD MUNTINS

⊄ OF BUILDING

WOOD DOORWAY ETC.

WOOD SILL

MOLDED BRICK

STONE STEPS

GRADE LINE

MEAS + DRAWN Kenneth Clark 1926

COURT · HOUSE · MT · HOLLY · N·J·
· THREE · EIGHTH · INCH · SCALE · DETAIL · FRONT · ELEVATION ·

Detail of Front Elevation
BURLINGTON COUNTY COURT HOUSE, MOUNT HOLLY, NEW JERSEY

Detail of Front Entrance Doorway
BURLINGTON COUNTY COURT HOUSE, MOUNT HOLLY, NEW JERSEY

Detail of Rear Entrance Doorway
BURLINGTON COUNTY COURT HOUSE, MOUNT HOLLY, NEW JERSEY

Detail of Main Cornice
BURLINGTON COUNTY COURT HOUSE, MOUNT HOLLY, NEW JERSEY

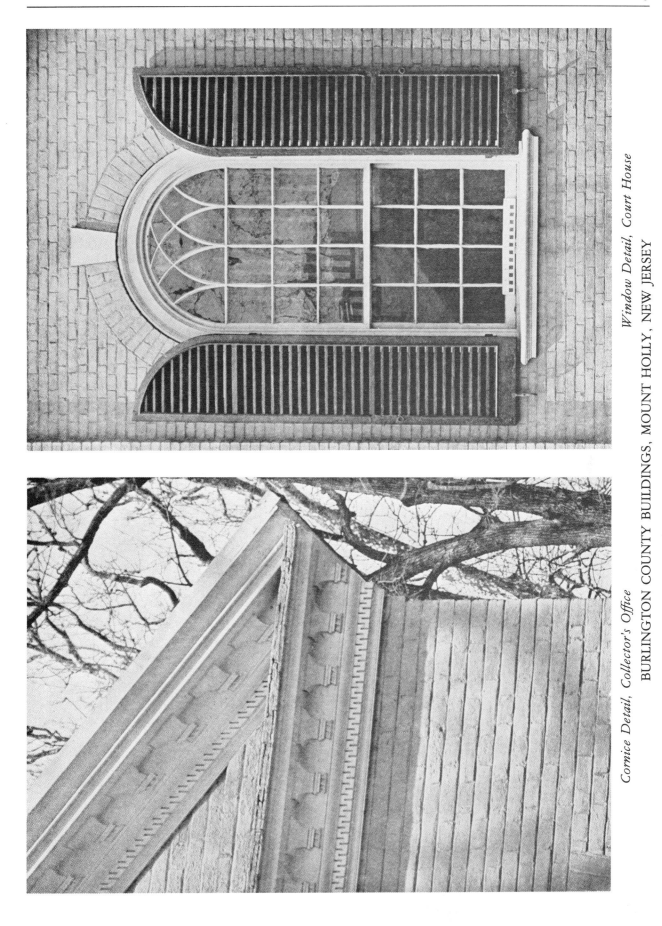

Window Detail, Court House

Cornice Detail, Collector's Office
BURLINGTON COUNTY BUILDINGS, MOUNT HOLLY, NEW JERSEY

Detail of Cupola
BURLINGTON COUNTY, COURT HOUSE, MOUNT HOLLY, NEW JERSEY

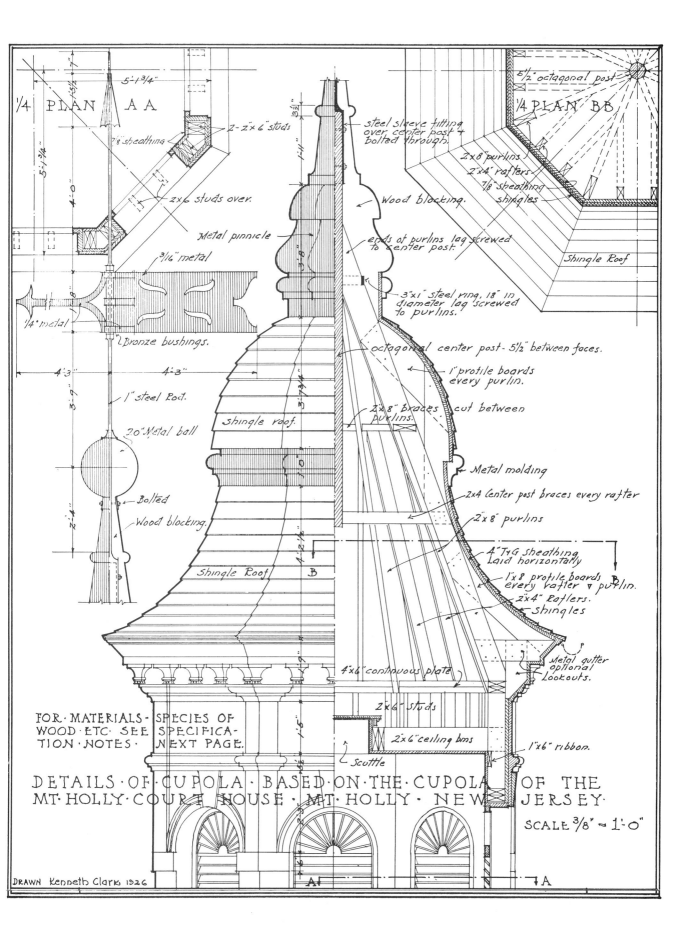

¼ PLAN AA

5'-1¾"

5'-3¼"

4'-0"

⅞ sheathing

2-2"x6" studs

2x6 studs over.

Metal pinnacle

3/16" metal

¼" metal

Bronze bushings.

4'-3"   4'-3"

3'-9"

1" steel Rod.

20" Metal ball

Bolted

Wood blocking.

steel sleeve fitting over center post & bolted through.

Wood blocking.

ends of purlins lag screwed to center post.

3"x1" steel ring, 18" in diameter lag screwed to purlins.

octagonal center post - 5½" between faces.

1" profile boards every purlin.

2"x8" braces cut between purlins.

Metal molding

2x4 center post braces every rafter

2"x8" purlins

4" T&G sheathing laid horizontally

1"x8" profile boards every rafter & purlin.

2"x4" Rafters.

shingles

Metal gutter optional Lookouts.

¼ PLAN BB

5½" octagonal post

2"x8" purlins
2"x4" rafters
⅞ sheathing
shingles

Shingle Roof

Shingle roof.

Shingle Roof

B

Shingle roof.

B

4"x6" continuous plate

2"x6" studs

2"x6" ceiling bms

Scuttle

1"x6" ribbon.

FOR·MATERIALS·SPECIES·OF
WOOD·ETC·SEE·SPECIFICA-
TION·NOTES·NEXT PAGE.

DETAILS·OF·CUPOLA·BASED·ON·THE·CUPOLA·OF·THE
MT·HOLLY·COURT·HOUSE·MT·HOLLY·NEW·JERSEY.

SCALE ⅜" = 1'-0"

DRAWN Kenneth Clark 1926

A   A

Detail of End Elevation, Collector's Office
BURLINGTON COUNTY BUILDINGS, MOUNT HOLLY, NEW JERSEY

Lower Delaware Valley Architecture

Text by
Jewett A. Grosvenor
Photographs by
Kenneth Clark and PH. B. Wallace
Originally published in 1920 as White Pine Monograph
Volume VI, Number 3

Detail of Façade
HOUSE AT GREENWICH, NEW JERSEY

THE WOODEN ARCHITECTURE OF THE LOWER DELAWARE VALLEY

THE Massachusetts colonist disregarded the abundant supply of stone about him and built a timber house. The early Pennsylvania colonist, hailing from a different part of England, settled in a land heavily wooded with a plentiful supply of the best timber heart could wish and used it merely to construct a log cabin for temporary shelter until he had time to quarry stone or bake bricks and build a dwelling of a type like that to which he had been accustomed in the mother country. If one may be permitted the indulgence of making a very bromidic observation, we are all creatures of habit. In no one particular is our addiction to hereditary custom more likely to come to the surface than in matters of architecture. This tendency on the part of the first settlers to stick to their own several architectural traditions has been pointed out more than once.

Although the persistent ignoring of physical conditions and clinging to traditional preference for materials and methods of construction, which the colonists, their fathers, and their grandfathers before them had been used to in England, gave the domestic architecture of our earliest Colonial period both variety and a pronounced individual bias, according to the town or shire the settlers had come from, common sense and necessity in time brought modifications, while independence of action and originality grew apace. Independent action, however, in the face of customary usage was always somewhat of an exception; and as exceptions are generally of special interest, for their comparative rarity if for no other reason, so we find it in the case of the wooden houses of Eastern Pennsylvania, West Jersey and Delaware, a portion of the land where

the majority of the English settlers showed their traditional preference for stone or brick.

The Swedes in Delaware apparently had no predisposition against timber and used it. Among the colonists of British origin, the men of West Jersey, notwithstanding the excellent early brickwork there to be seen, were the first to adapt themselves to conditions with good grace, make a virtue of necessity, and build of timber when it was well-nigh impossible to get stone and nearly as difficult to come by suitable brick. Their soil was stoneless, good brick clay was scarce, the pine growth was abundant, and they did the obvious thing—they built of timber. And posterity has never had cause to regret their choice. In Pennsylvania wooden structures of any amenity came later—the end of the eighteenth century and the early part of the nineteenth—and reflected the characteristics of the time. In each of these three states, the domestic wooden architecture has peculiarities of its own, but all of it yields interest and from all of it something suitable for modern adaptation can be gained.

In Delaware, at a very early date, dwellings of the type of the oldest house in Dover—chosen as an illustration, not for appearances, but for its archaeological value—were not uncommon and were also to be seen in the Swedish portions of Philadelphia. They were of mixed English and Swedish parentage. The outstanding chimney is English, the gambrel roof with its sharp lower pitch sounds a Scandinavian note in contour. The type is simple but strong, and susceptible of interesting development. The old batten shutters, with boards set chevron-wise to form a herringbone figure, still left on one of the lower windows, are to be noted as characteristic of this part of the country. Despite the neglect and ill usage to which this house has plainly been subjected, its clapboard walls and shingle roof are still staunch and weather-worthy.

Across the Delaware River, in South and West Jersey, where the easy and substantial affluence of a fertile farming region of large plantations encouraged building, one finds a different condition obtaining. From Salem up to Burlington or Bordentown, in the face of stone and brick tradition and the precedent of numerous fine examples of early brickwork, especially in the neighborhood of Salem, many of the prosperous farmers soon took to the course of least resistance and built of wood.

One of the first, and one of the most interest-

OLDEST HOUSE AT DOVER, DELAWARE

ing, examples of West Jersey wooden architecture is The Willows, on a point of land jutting into Newton Creek near Gloucester, a structure dating from about 1720. It was once a handsome country seat, but, years ago, owing to the encroachment of manufacturing plants, became untenable as a residence and was abandoned to tenancy and truck-farming. Nevertheless, despite its external dilapidation and sorry surroundings, the house presents features that the student of architecture cannot afford to neglect. Indeed, just because of its dilapidation, some of its structural peculiarities have become visible and admit of easy analysis in a way that would be impossible in a structure kept in decent repair. Besides being one of the earliest wooden houses, it shows the combination of a later addi-

THE WILLOWS—1720—GLOUCESTER, NEW JERSEY

HOUSE AT GREENWICH, NEW JERSEY

tion, with its opportunity for making comparisons, not to be found in any other of the contemporary buildings in the neighborhood. The older or eastern portion (to the right in the picture) is built of three-inch white pine planks, double grooved with sliding tongues and even joints dovetailed together at the corners. The structure is really a piece of cabinet work rather than a piece of carpentry, and is a monument to the skill of the joiner—the old term is peculiarly ap-

Of an entirely different type are the capacious, foursquare, clapboarded houses, of slightly later date, that are to be found aplenty throughout West Jersey. Of this class the house at Bordentown may be regarded as representative, or the house at Salem. These houses boasted a symmetrical, rectangular plan with central hallway from front to back,—rooms on each side of it, an ell extension at the rear, and chimneys at each gabled end. There is rarely any attempt at em-

HOUSE—1740—BORDENTOWN, NEW JERSEY

propriate for the artisan in this instance—who framed it together. It is only since the loosening and dropping off of the corner boards that this feature of construction has become visible.

The Willows, as are also nearly all other old West Jersey wooden houses, is "brick-paned," or lined with a solid brick wall inside the plank or clapboard exterior and between the studs. So substantial is the structure and so thorough the workmanship, that, after nearly two centuries, only slight repairs and reasonable care are needed to make it as fit as it ever was.

bellishment, for most of these houses were built by plain Friends who had conscientious objections both to collars on their coats and ornament on their dwellings. The simplest kind of cornice is ordinarily the only concession to the impulse for decoration. Otherwise, nine out of every ten are as plain as the proverbial pipestem, but their proportions are usually agreeable and their general aspect seems to fit in with the quiet affluence and unassuming thrift that furnishes forth their old mahogany midday dinner table with blue Canton and old silver and yet

Doorway Detail
HOUSE AT BORDENTOWN, NEW JERSEY

EWING HOUSE—1800—MOORESTOWN, NEW JERSEY

BILDERBECK HOUSE — 1813 — SALEM, NEW JERSEY

sets master in shirt sleeves and men in overalls side by side to devour the plenteous fare.

Of more urbane and polished type by far, is the Haddonfield house that appears in the illustration. Chronologically characteristic of the early nineteenth century, it also combines in its aspect an unmistakable note of Quaker reticence and austerity. The usual Classic Revival type is perfectly familiar, but here is a Classic Revival type pared down, attenuated, robbed of all self-assertion, and compressed into Quaker simplicity. The residuum from the transformation

a new building of exceedingly restrained and austere design: "That ain't no architecture; that's a packing box."

Of great charm is the house on the Haddonfield Pike, set amid its box bushes and ancient yew trees, with its modest porches and its side wing expanded into a broad gambrel-roofed structure, only a little less in size than the main body of the dwelling. The house is thoroughly representative of the town, which is itself representative of the best traditions of West Jersey wooden architecture, peculiarly reminiscent of Elizabeth

HOUSE — 1810 — HADDONFIELD, NEW JERSEY

turns out to be singularly agreeable. The house, with a door at one side of the front, two windows beside it, three windows on the front of the second floor, and a wing or extension at the side or the rear, belongs to a well recognized type that flourished at the beginning of the nineteenth century. But this type commonly exhibited an accompaniment of some emphatic Classic details. Here, on the contrary, we have the Classic items reduced to the lowest terms, like a rigid Quaker's speech and fashion of garb, and with some elegance withal. Had any more been subtracted, there would have been a risk of meriting the shrewd old countryman's criticism, upon seeing

Haddon, that firm and virile-minded seventeenth-century maiden who assumed her father's interests, founded the town, courted and married — of her own initiative, tradition says — and continued to sign and be known by her maiden name.

Another house typical of West Jersey domestic architecture in wood, the previously mentioned eighteenth-century building at Bordentown, might be called the decorated member of the symmetrically arranged rectangular dwelling class. The detail of the Bordentown building (1740) is rather unique in point of the course of small panels below a frieze ornamented with

HOUSE NEAR PHILADELPHIA, PENNSYLVANIA

drapery swags, the fine mouldings of the window casings, and the slender, semi-engaged pillars of the door frame that suggest the work of an artisan from the Dutch counties of North Jersey.

A reversion to the old type of smooth-jointed, grooved-plank construction may be seen in the Moorestown house, dating from about 1800. Both in plan and architectural amenity the illustrations show this building is a highly creditable

coast towns or the inland Jersey towns but a little way from New York to be overpowered with the dreary horrors perpetrated anywhere between 1860 and 1885, or even later. Between those years the jig-saw decorator was rampaging at large and embellishing (?) the wooden packing boxes that prostituted a noble building material and did more to give wood, for the time being, a bad name as an architectural medium than any other one thing in the history of build-

HOUSE AT WOODSTOWN, NEW JERSEY

exemplar of what may be achieved in a wooden medium.

These several types of West and South Jersey wooden houses have set a precedent that has been assiduously followed by later generations in New Jersey towns, so far as material alone is concerned. How much better they might have followed or adapted it in the matter of architectural expression, the "man who was blind in one eye and couldn't see with the other" might tell at a glance. One needs only go through the

ing. The old houses show what charm frame dwellings were capable of presenting in intelligent hands.

The wooden architecture of the Lower Delaware Valley, while not so abundant as in some other parts of the country, for reasons already mentioned, is nevertheless invested with the merit of a distinct individuality, or several individualities, and has its share to contribute both to the story of house building in America and to modern inspiration.

BILDERBECK HOUSE

17 MARKET STREET

DOORWAYS AT SALEM, NEW JERSEY

HOUSE AT 17 MARKET STREET, SALEM, NEW JERSEY

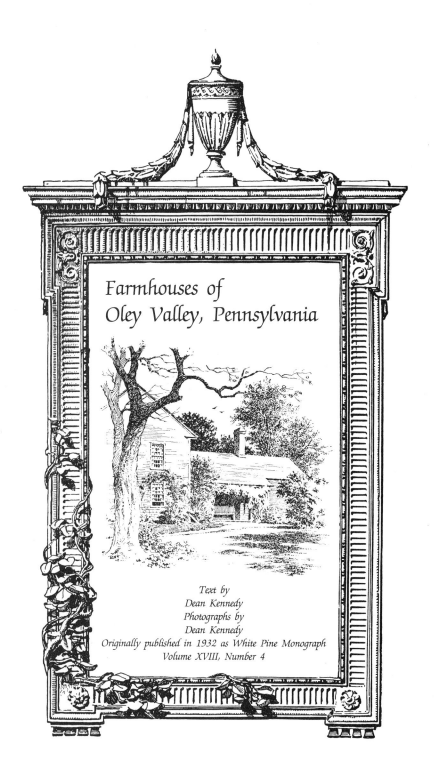

Farmhouses of
Oley Valley, Pennsylvania

Text by
Dean Kennedy
Photographs by
Dean Kennedy
Originally published in 1932 as White Pine Monograph
Volume XVIII, Number 4

Doorway Detail
HUNTER HOUSE, NEAR YELLOW HOUSE, PENNSYLVANIA

FARMHOUSES OF OLEY VALLEY
BERKS COUNTY, PENNSYLVANIA

THE fertile lime-stone valley of Oley in the county of Berks, Pennsylvania, furnishes us with a valuable architectural heritage. The area of the valley consists of some thirty square miles and is situated near Reading in the eastern portion of the county as may be seen in the accompanying map. In Friedensburg (Oley), Pleasantville, Lime Kiln, Spangsville, and Yellow House we find excellent examples of century old farmhouses which contribute much to our understanding of the homes of our ancestors and to the records of early American architecture.

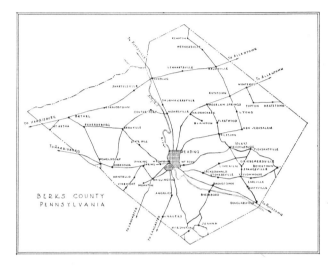

The settlers who arrived during the latter part of the seventeenth and the early eighteenth centuries were chiefly French Huguenot refugees, fleeing from their native country to escape the persecutions of the times and coming in search of religious freedom in a place where they could make their new homes.

William Penn, when first laying out his colony in 1682, divided it into three counties, Philadelphia, Chester, and Bucks, all radiating from a point where the city of Philadelphia now stands. Naturally, the first settlers followed the water courses to their best advantage. The Swedes, coming up the Schuylkill River, settled above the mouth of the Monocacy Creek, where the town of Douglassville is now located. There today may be found some of the earliest signs of civilized life in the county, and a few of the early houses, including the Mounce Jones House, erected in 1716, which is perhaps the oldest house still standing in what is now Berks County.

Following the Swedes, the English Quakers, coming up the Manatawny Creek, located in the present Oley, along with a number of French and Swiss Huguenots, as well as German families. Such Huguenot names as DeTurk, De LaVan, De LaPlank, Bertelot, De-Long, as well as others have survived from the earliest

settlement to the present time.

In 1698, John Keim, a young German, was the first known settler to stake and lay claim to a tract of land in Oley. In 1706, returning from Europe with his bride, he began to clear his land for cultivation and later erected a stone dwelling, which is still standing. In 1712, Isaac DeTurck, a brother-in-law of Keim, coming from New York, settled on a tract near the present village of Friedensburg.

Between this time and 1720 other French Huguenots, as well as the Lees, English Quakers, headed by Anthony Lee, the first to arrive in Oley, in 1718 settled in what is now known as the village of Pleasantville. The Lees were soon joined by the Boone family and others of the same religion, and as early as 1726 they had organized themselves into a separate congregation and built their first church of logs. In 1736 it is known that George Boone collected funds for a larger and better structure. Later this also was found to be insufficient, and the third, the present Exeter Friends' Meeting House, was built sometime before 1800, the exact date is unknown. It is one of the oldest Quaker Meeting Houses in Pennsylvania, outside of Philadelphia.

In the possession of Daniel Fisher, present owner, and great-grandson of Henry Fisher, the builder of the Fisher Homestead near Yellow House, there is a brief of title stating that on April 20, 1682, during the reign of King Charles II, William Penn, of Worminghurst, Sussex County, England, deeded to John Sheiras of York County, England, 1,000 acres of land in the Province of Pennsylvania. The land changed hands several times until, in 1791, John Lesher sold some three hundred acres to Henry Fisher, showing that a large part of the land must have been sold off before this time. The buildings on the land were already old and

EXETER FRIENDS' MEETING HOUSE, NEAR STONERSVILLE, PENNSYLVANIA

as soon as it could be arranged, preparations were made for building a new house, which was finished in 1801. Nearby stands a large springhouse in which, no doubt, the family lived while the house was being erected.

The Kaufman House, illustrated on page 118, is an example of one of the original houses to be found in the valley although somewhat larger than the majority. This house, like so many others in the neighborhood, is now used as an outbuilding for the new house which was built when larger quarters were needed. These first houses are sturdy old places of stone and timber, entirely lacking in embellishment but beautiful in their proportions. Often built into the slope of a hill near or directly over a spring, the houses thereby served the double purpose of dwelling and springhouse. The plan is usually rectangular with but one or two main rooms on each floor. The kitchen is on the lower floor, in cases where the house was built into a hill. Here one may see the huge fireplace with smoked hewn timber lintel and simple board mantel shelf.

The Fisher House, today, is perhaps the best preserved in the Oley Valley, outside of some that have been restored, and may well be considered typical of

the Pennsylvania farmhouse. The exterior walls are of limestone, as are the majority of the dwellings of this type. There is a pedimented doorway at the center of the broad front of the main rectangle, opening into a spacious hallway, dividing the large rooms devoted to living and dining purposes. The kitchen is in a wing which projects from the main house. This addition is also built of rough stone but the original detail has been modified.

The flat arch above the window openings deserves mention, because a similar form of stone arch frequently occurs in houses throughout Pennsylvania. Here in the Fisher House a flat arch is fashioned of wood, with a central key block of greater height than the adjoining pieces. The cornice mouldings and doorway detail seem to be identical with many of the other houses and it is interesting to note how the modification of the same detail has been used on the main cornice, the pedimented doorway, the mantels, and the cornices of rooms and hallways. The detail may have been original with the builder or more probably copied from some carpenter's manual so much in use at that time. The chief carpenter of the Fisher House is known to have been Gottlieb Drexel, and to him belongs the credit of the fine paneling, stairway, friezes, cornices, and other architectural features.

The house contains six fireplaces, some of which are faced on the outside with Italian marble and plastered on the inside. Possibly the most beautiful of these, and the most intricate in its detail, is the one in the bridal chamber or guest room, illustrated on pages 111 and 112.

The George Boone House, erected in 1733, with its whitewashed stone walls, is interesting in its spacing of windows on the front wall and the lack of windows and the mere slits which occur on the end walls. A line of timber may be seen projecting on the end walls, as if to form the gable end but on further thought the wall was moved back some few feet. Evidently the wall was never terminated to form this gable as no distinct jointing can be seen where the new wall would have joined the old, and the date 1733 scratched into one of the sandstone quoins of the rear wall leads one to believe this was the original

wall. However, the difference in the pitch of the two roofs gives anything but a pleasing proportion. It is said that George Boone, the grandfather of Daniel Boone, was well content to live in his simple log cabin nearby until his death in 1744, declaring the new home was much too pretentious for his simple tastes. It was actually occupied by his eldest son, George Boone II.

It is an interesting fact that these early homesteads, found within a radius of a few miles, have remained in the possession of descendants for the past century.

Much could and has been said of the picturesque outbuildings, the spring and tenant houses, bake ovens and smokehouses spotting the Oley countryside, with their whitewashed or plastered stone walls, and many still retaining their weathered red tile roofs. All of these, worked out and planned to the best advantage, together with the well built farmhouses, speak loudly of the thrift, domestic loyalty, and good taste of these early immigrants.

GEORGE BOONE HOUSE, NEAR LIME KILN, PENNSYLVANIA

Doorway Detail
KNABB HOUSE, NEAR LIME KILN, PENNSYLVANIA

Doorway Detail
FISHER HOUSE, NEAR YELLOW HOUSE, PENNSYLVANIA

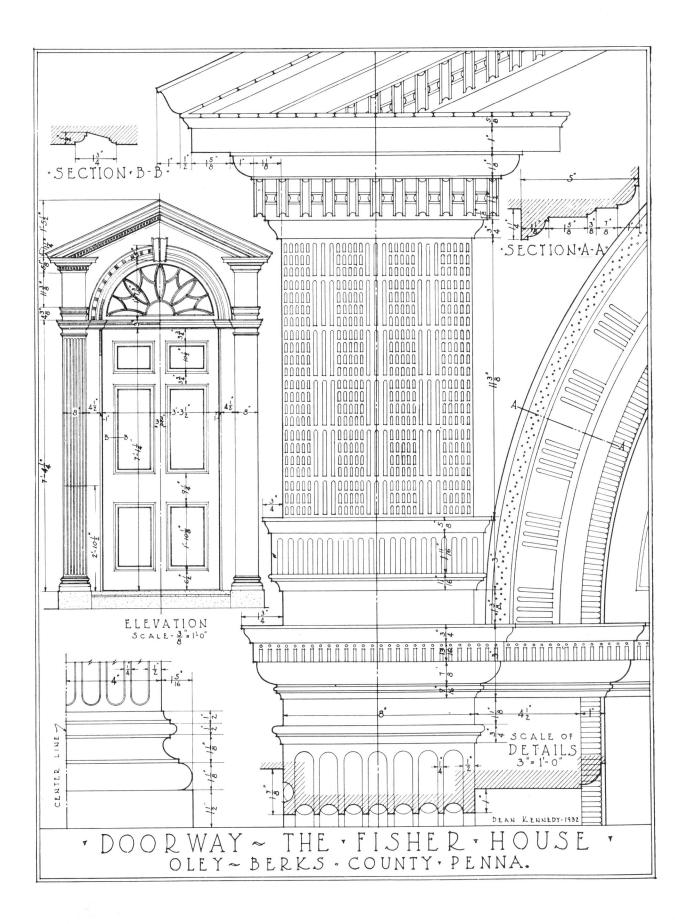

·SECTION·B-B·

·SECTION·A-A·

A

ELEVATION
SCALE - 3/8" = 1'-0"

CENTER LINE

SCALE OF
DETAILS
3" = 1'-0"

DEAN KENNEDY·1932

· DOORWAY ~ THE · FISHER · HOUSE ·
OLEY ~ BERKS · COUNTY · PENNA.

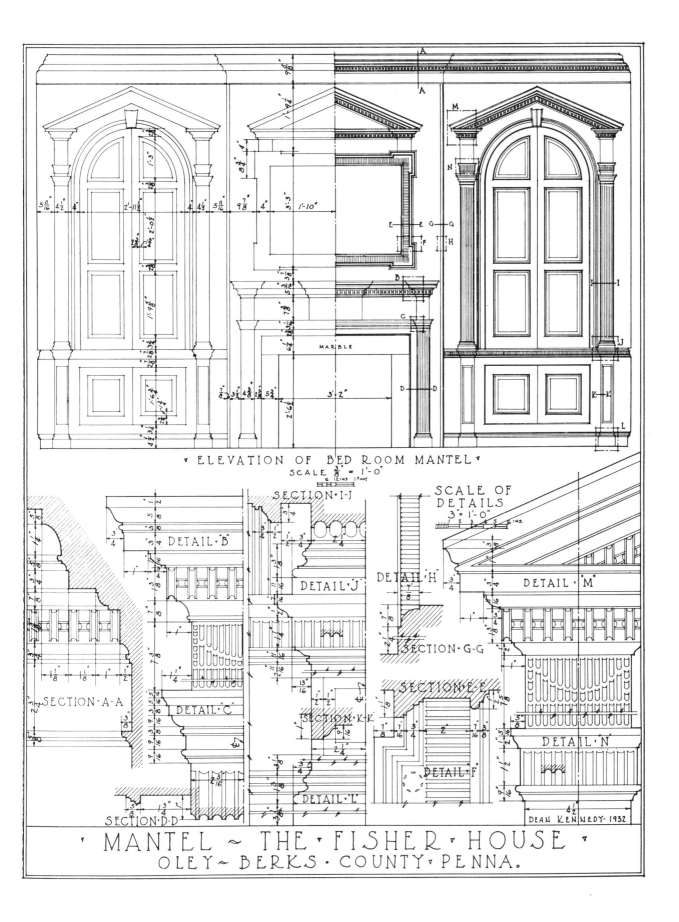

· ELEVATION OF BED ROOM MANTEL ·
SCALE ⅜" = 1'-0"

MARBLE

SECTION·I·J

DETAIL·B

DETAIL·J

DETAIL·H

SCALE OF
DETAILS
3" 1'-0"

DETAIL·M

SECTION·G·G

SECTION·A·A

DETAIL·C

SECTION·K·K

SECTION·E·E

DETAIL·N

DETAIL·F

SECTION·D·D

DETAIL·L

DEAN KENNEDY · 1932

· MANTEL ~ THE · FISHER · HOUSE ·
OLEY ~ BERKS · COUNTY · PENNA.

Bridal Chamber
FISHER HOUSE, NEAR YELLOW HOUSE, PENNSYLVANIA

FISHER HOUSE, NEAR YELLOW HOUSE, PENNSYLVANIA

First Floor Hallway
FISHER HOUSE, NEAR YELLOW HOUSE, PENNSYLVANIA

Doorway Detail
SPANG HOUSE, SPANGSVILLE, PENNSYLVANIA

SPANG HOUSE, SPANGSVILLE, PENNSYLVANIA

HUNTER HOUSE, NEAR YELLOW HOUSE, PENNSYLVANIA

ORIGINAL KAUFMAN HOUSE, NEAR PLEASANTVILLE, PENNSYLVANIA

Bethlehem, Pennsylvania

Text by
Karl H. Snyder
Photographs by
Kenneth Clark
Originally published in 1927 as White Pine Monograph
Volume XIII, Number 4

OLD MORAVIAN SEMINARY (BELL HOUSE)—1745–1746—BETHLEHEM, PENNSYLVANIA

MORAVIAN ARCHITECTURE
OF BETHLEHEM, PENNSYLVANIA

THE colonial domain in America comprised regions which differed conspicuously from one another in climate, soil, and economic opportunity. The races which came to dwell in these new lands were no less diverse than the country.

By 1763, the total white population of the region from Maine to Georgia was not far from 1,250,000. It has been estimated that more than one-third of the inhabitants were newcomers, not of the stock of the original settlers. These newcomers were chiefly French, German, and Scotch-Irish, but their influence on colonial life was less than their numbers might suggest. The Germans, as farmers, contributed greatly to the prosperity of the communities where they cultivated their lands.

Pennsylvania, the last of the colonies to be founded, except Georgia, contained a kaleidoscopic collection of people of different bloods and religions. The population increased from 50,000 in 1730 to more than 200,000 in 1763 due, in the largest part, to the thousands of Scotch-Irish and Germans, who poured into the colony, The Scotch-Irish went to the region of the Susquehanna. The Germans usually settled near the old counties, where they could devote themselves to the cultivation of the soil, and to the maintenance of their many peculiarities of life and faith. The life of these Germans, Moravians, Mennonites, Schwenkfelders, Dunkards, and many, many others, was marked by simplicity, docility, mystical faith, and rigid economy. The Germans were unaccustomed to liberty of thought as to political liberty, and it produced a new sect or religious distinction almost every day. Some of them were inclined to monastic and hermit life.

Most of the German sects left the Quakers in undisturbed possession of Philadelphia, and spread out into the surrounding region which was then a wilderness. They settled in a half-circle beginning at Easton-on-the-Delaware passing up the Lehigh Valley into Lancaster County and down the Susquehanna.

The Moravians are a Christian sect founded by disciples of John Huss during the fifteenth century in Moravia, a former province in the northwest of Austria-Hungary, and now included in Czechoslovakia. The sect is also known as the United Brethren or Unity of Brethren. The Moravians, under the patronage of Count Zinzendorf, were constantly seeking a wider field for their missionary work, and we find that as early as 1727 Count Zinzendorf, who later became the leading bishop of the Moravian Church, purchased a tract of land in the province of Georgia for a colony of Schwenkfeldian exiles from Silesia. It was not until the spring of 1735, however, that the Moravians made their settlement in Georgia, and brought the gospel to the Indians and the negro slaves. A school for Indian children was established on the Savannah River, a mile above the town of Savannah. The war between England and Spain, which occurred a few years after the settlement in Georgia, interfered with the work of the missionaries, so that in 1740 they decided to move their activities to Pennsylvania where they could be among the German colonists, and administer to the spiritual welfare of the Indians in that region. It is interesting to note that the Indians near Bethlehem, who had been converted to Christianity by the Moravians were the only ones who did not follow Pontiac when in 1763 the tribes swept eastward into Pennsylvania, burning, murdering, leveling every habitation to the ground.

In the "Forks of the Delaware" they purchased five thousand acres of land, and at Nazareth in May, 1740, founded their first settlement in Pennsylvania, under the leadership of Rev. George Whitefield. The first house that was built, now called the Grey House, is illustrated on page 122. It is a log house constructed of hewn white oak. The most imposing building erected by the Moravians at Nazareth was intended originally to be the manor house for Count Zinzendorf. The style and construction are typical of Moravian colonial architecture. The builders evidently were striving to create the domestic atmosphere of the houses in Silesia in order to make Count Zinzendorf feel at home in the new country. The cornerstone was laid in 1755 and the structure completed in 1759. The building was never occupied as a residence, but was formally opened in 1759 as the Boys' School of the American Province of the Church. The cupola was added in 1785. This may be seen in the illustrations on pages 125 and 126.

The main body of the Moravians did not remain long in Nazareth, due to differences between Rev. Whitefield and his followers. They decided to purchase five hundred acres of land which lay at the junction of the Lehigh River and the Monocacy Creek. Here they were reinforced by colonists who had come to the "back country" from Germantown, Pennsylvania, and a new settlement was established. They built their first log house in December, 1740, in what is now the thriving industrial city of Bethlehem. It was Count Zinzendorf who, stimulated by the associations connected with the celebration of Christmas, gave the place its Biblical name.

Of the several Moravian buildings at Bethlehem which remain in but slightly altered condition, the ⊔-shaped group made up of the Brothers' House, the Sisters' Home, and the Seminary are the most exotic, and show a well-defined architecture of German derivation. We are reminded, by the heavy stone and timber construction, the steep roofs with two rows of sloping dormers, and the flanking buttresses, of the medieval buildings of the old world.

The left wing of the group is the second house that was built in Bethlehem. During the first years of the settlement, "The House on the Lehigh" served as a home and hospice, manse and church, administration office, academy, dispensary and town hall; the loved resting place of weary pilgrims, the busiest center to be found far and wide, sought out by the inquisitive and expatiated on by many a gossip, who told wonderful

GREY HOUSE, NAZARETH, PENNSYLVANIA
The First Moravian House in Pennsylvania

stories about the Moravians. The building was called *Gemeinhaus* in the German nomenclature or community house. It stands today in its original form although its massive logs are hidden by clapboards.

Our frontispiece illustration shows the middle building of the group, called in local parlance, the Bell House, but officially known as the old Seminary among the Moravian properties because the boarding school for girls occupied it from 1749 to 1790. It was built originally to contain the refectory of the single men, a general dining room connected with the Community House, and the quarters for the married people. The foundation lines were staked off, August 24, 1745, but the building was not completed until October of the following year. The bell turret, which was to contain the first town clock, was erected in June, 1746. Augustine Neisser of Germantown, began work on the clock in April, 1746, but did not complete it until February 15, 1747. The three bells, one large and two smaller ones, were cast by Samuel Powell, who was also the first innkeeper on the south side of Bethlehem. The weather vane, which still surmounts the little turret was made from a drawing by Cammerhoff. The historic emblem of the Church, a lamb with a banner, is part of the design.

The Moravian Sisters' House forms the right wing of this group. The designers appear to have harked back to the Gothic architecture of the smaller German towns with which they were familiar. They seem to have felt the need of the huge buttresses even though there are no vaults or cross arches to sustain.

The court yard, or open square formed by the three buildings, was long the scene of Holy-Day musicals, and Harvest-home festivals.

MORAVIAN SISTERS' HOUSE, BETHLEHEM, PENNSYLVANIA

Window Detail
SHIMER HOUSE, NEAR BETHLEHEM

The region around Bethlehem is dotted with the typical unpretentious Pennsylvania farmhouses, the original dwellings of Moravian farmers. The outside walls are of stone with gables at the lateral ends. It will be noted that the windows have fifteen lights, nine in the upper sash and six in the lower. The Steuben House at Republic Hill (page 137), is one of the exceptions to the usual spacing of two windows on each side of the door opening. Paneled shutters, exposing a plain flush surface, on the outside when closed, and the panels on the inside when opened, are typical. In the examples illustrated, the rail between the two large panels seem arranged so as to continue the line of the meeting rail of the window.

The measured drawings of the doorway of the Freeman House in Freemansburg, about two miles from Bethlehem, show an important piece of work, important because it contains some odd twists to well known architectural motives. The tulip decoration recalls the Pennsylvania Dutch painted furniture, but the cornice, frieze, entablature, and columns were not taken out of the "Books" or *Builders' Assistants.*

The term "Pennsylvania Dutch," used in speaking about the eighteenth-century buildings in Pennsylvania, has come to mean German and Swedish, more often than Netherlandish. It was the German frontiersmen who played the largest part in the back country around Bethlehem, and their influence on the adopted style of building has given the domestic architecture of the state an air of its own, quite different from that found in the other colonies. The feeling that pervades the old farmhouses has been so imbided and availed of by modern architects that there are many thousand just such houses everywhere throughout this territory and it is difficult for the traveler to believe in the real antiquity of some of the old houses, their flavor is so very modern.

In these days of cold facts, commercialism, and expensive frontages, one feels that Art is being shoved into the discard, and when a good old building is demolished to make room for an "Ultra Modern" duplex apartment house, one knows that the Goddess of Beauty is getting an awfully raw deal. It is remarkable and commendable, that the Moravians in Bethlehem, have preserved, practically without an alteration, their group of institutional buildings in the heart of an industrial city.

Detail of Stone Wall
STEUBEN HOUSE, REPUBLIC HILL

MANOR HOUSE FOR COUNT ZINZENDORF, NAZARETH, PENNSYLVANIA
Opened in 1759 as The Boys' School of the American Province of the Church

Cupola
MORAVIAN BOYS' SCHOOL (NAZARETH HALL)—1785—NAZARETH, PENNSYLVANIA

Cupola
MORAVIAN CHURCH—1803—BETHLEHEM, PENNSYLVANIA

Doorway
HOUSE ON MINSI TRAIL, BETHLEHEM, PENNSYLVANIA

Detail of Main Entrance
JOHN FREEMAN HOUSE, FREEMANSBURG, PENNSYLVANIA

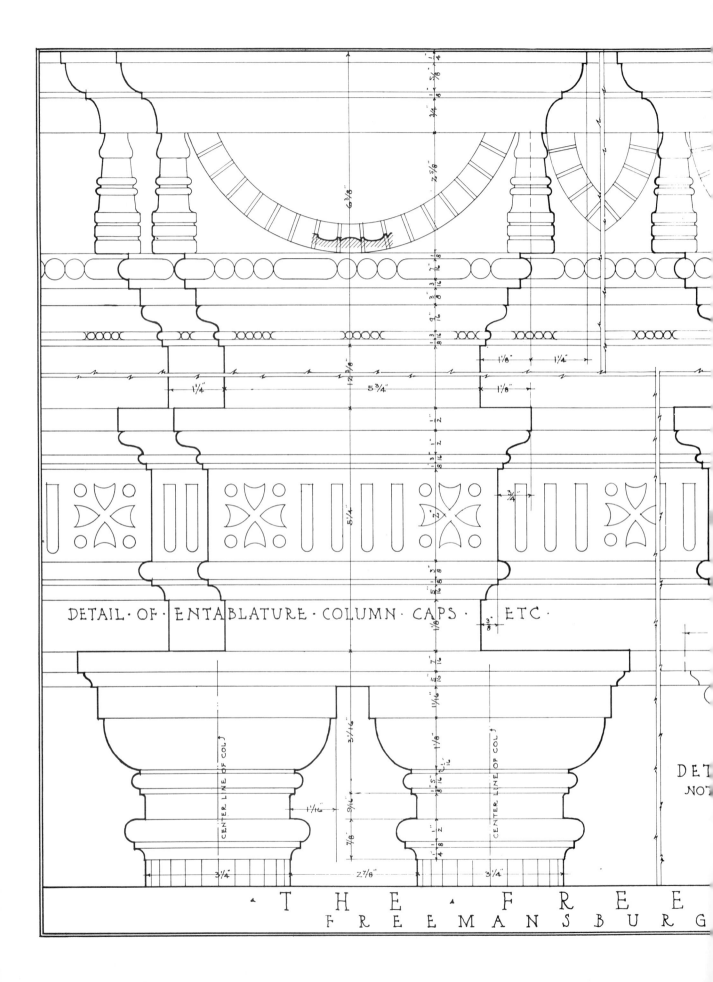

DETAIL · OF · ENTABLATURE · COLUMN · CAPS · ETC ·

CENTER LINE OF COL.

CENTER LINE OF COL.

· THE · FREE

FREEMANSBURG

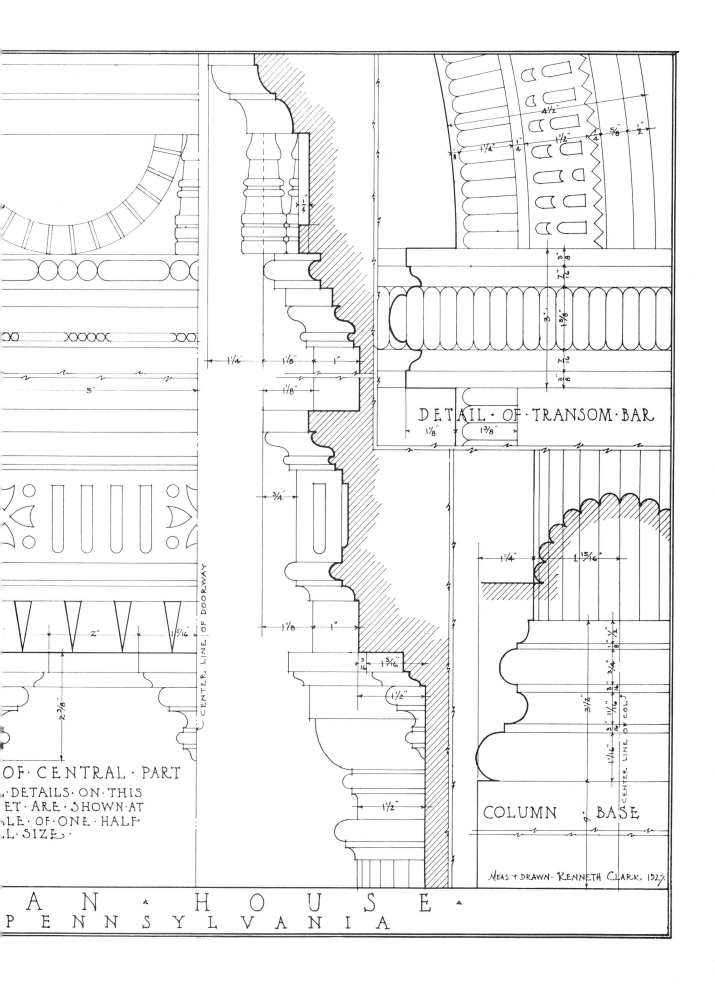

DETAIL · OF · TRANSOM · BAR

OF · CENTRAL · PART

· DETAILS · ON · THIS
ET · ARE · SHOWN · AT
LE · OF · ONE · HALF
L · SIZE ·

COLUMN · BASE

MEAS + DRAWN · KENNETH CLARK · 1927.

CENTER · LINE · OF · DOORWAY

CENTER · LINE · OF · COL.

AN · HOUSE ·

PENNSYLVANIA

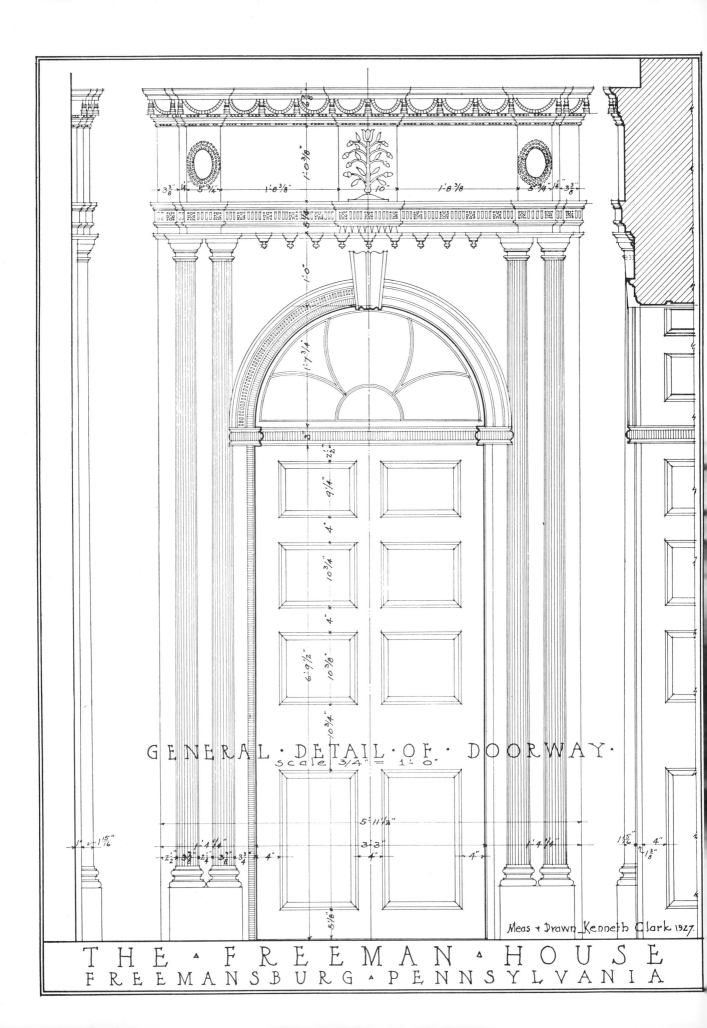

GENERAL · DETAIL · OF · DOORWAY ·
scale 3/4" = 1'-0"

Meas + Drawn Kenneth Clark 1927.

THE · FREEMAN · HOUSE
FREEMANSBURG · PENNSYLVANIA

Doorway
JOHN FREEMAN HOUSE, FREEMANSBURG, PENNSYLVANIA

ROHN FARMHOUSE, BATH VILLAGE NEAR BETHLEHEM, PENNSYLVANIA

SHIMER FARMHOUSE, ALONG LEHIGH RIVER NEAR BETHLEHEM, PENNSYLVANIA

SHIMER FARMHOUSE—1801—NEAR BETHLEHEM, PENNSYLVANIA

STEUBEN HOUSE, REPUBLIC HILL, PENNSYLVANIA

Shop Window Detail
HOUSE AT HECKSTOWN, PENNSYLVANIA

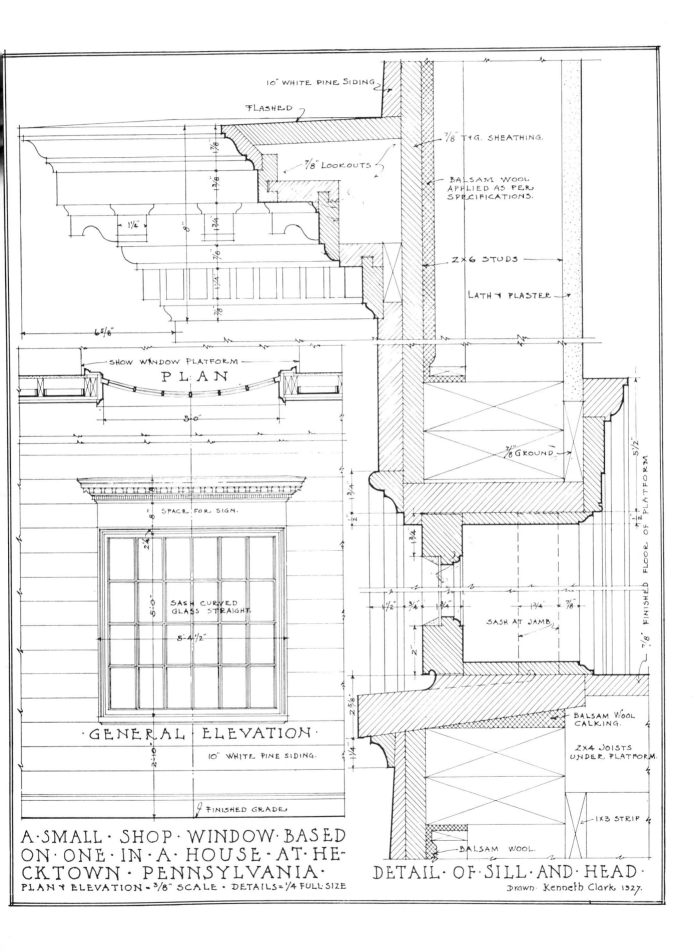

10" WHITE PINE SIDING

FLASHED

7/8" T&G. SHEATHING.

7/8" LOOKOUTS

BALSAM WOOL APPLIED AS PER SPECIFICATIONS.

2×6 STUDS

LATH & PLASTER

SHOW WINDOW PLATFORM

PLAN

7/8" GROUND

FINISHED FLOOR OF PLATFORM

SPACE FOR SIGN.

SASH CURVED GLASS STRAIGHT.

5-4 1/2"

SASH AT JAMB

·GENERAL·ELEVATION·

10" WHITE PINE SIDING.

BALSAM WOOL CALKING.

2×4 JOISTS UNDER PLATFORM.

FINISHED GRADE

1×3 STRIP

BALSAM WOOL.

A·SMALL·SHOP·WINDOW·BASED
ON·ONE·IN·A·HOUSE·AT·HE-
CKTOWN·PENNSYLVANIA·
PLAN & ELEVATION = 3/8" SCALE · DETAILS = 1/4 FULL SIZE

DETAIL·OF·SILL·AND·HEAD·

Drawn· Kenneth Clark 1927.

Doorway
SHIMER FARMHOUSE, NEAR BETHLEHEM, PENNSYLVANIA

New Castle, Delaware

Text by
William C. Foster
Photographs by
Kenneth Clark
Originally published in 1926 as White Pine Monograph
Volume XII, Number 1

Main Façade
AMSTEL HOUSE — 1730 — NEW CASTLE, DELAWARE

NEW CASTLE, DELAWARE:
AN EIGHTEENTH-CENTURY TOWN

THERE are few communities today which have re-tained their early American flavor as completely as has New Castle, Delaware. The examples of our Col-onial architecture in such centers as Boston and Philadelphia and even in Baltimore are so surrounded by present-day business, or lie isolated without any sur-roundings, that they can give little of the feeling of the actual community which existed when they were built. Quebec, Salem, Charleston and New Orleans do pre-sent, each in certain quarters, this sense of the com-pleteness of the community, but New Castle, being a small town, presents the architecture of the middle col-onies even more completely than do these other cities of their respective sections.

Though small now and comparatively little known, New Castle was up to the early part of the nineteenth century quite an important place and her commerce brought considerable wealth to her citizens. Her his-tory had gone back to the earliest settlements, to the time of New Sweden and New Netherland. In fact, at the time Henry Hudson discovered what he termed the North River—our Hudson River—he also sailed up the South River—today the Delaware—and it was on that discovery that the Dutch, for whom Hudson was navigating, based their claims of ownership. How-ever, they were slow to act and the Swedes were the first to establish a colony on the banks of the South River, calling the district New Sweden.

The Dutch did not look upon this settlement with much favor and after the Swedes had been established some eleven years Governor Peter Stuyvesant of New Netherland, in 1651, acting partly on instructions from Holland and largely in accordance with his own vig-orous instincts, decided to take active measures to pro-tect the Dutch claims. Accordingly he proceeded per-sonally to New Sweden and established Fort Casimir, very near the site of the present town. There followed various turns of fortune for the fort. The Swedes cap-tured it in 1654, but the Dutch came back and in 1655 gained all of New Sweden, or Delaware, and renamed the settlement Fort Amstel, making it the seat of the Dutch government for the local colonies.

Disease and famine as well as a constant fear of the English caused most of the inhabitants to leave and in 1664 the English seized the whole district without much effort, again changing the name of the principal settle-ment, this time to New Castle. As New Castle it was frequently the meeting place of the legislature and later became the capital of the colony when Delaware was separated from Pennsylvania.

During the colonial days and the years of the Revo-lution the town played its part in making history. Two of the signers of the Declaration of Independence were residents, while a third had been born in New Castle.

And yet with this imposing background, as rich as that of many towns which are quite large today, New Castle at the end of the eighteenth century was as im-portant as it probably ever will be. The principal in-dustrial growth of the district is being assimilated by Wilmington, six miles to the north. While there are some factories with their resultant nondescript hous-ing, they are all grouped near the branch-line railroad which comes in to the west of the town proper. The compact older portion is still complete and removed from too much "progress."

Around the Common, which remains as an open park in the center of the town, are the public buildings. There is the old Court House, one wing of which is sup-posed to have been built about 1680, though the main portion dates from 1707. It was this building which served as a center for a twelve-mile radius which estab-lished the "northern boundary of the colonies on the Delaware," the arc which still is on the maps. Not far behind the Court House is the Episcopal Church, parts of which were built in 1701 and 1705.

To the east of the Court House is the very interest-ing square building (page 162), generally known as the Town Hall, but which has been put to various uses from time to time. Once it was the terminus of some sort of railroad and again it formed the end of a shed which covered the town market; at that time the fire engine was kept there and the upper floor was used as a town hall or common room. This three-story build-

Paneled Room End
AMSTEL HOUSE, NEW CASTLE, DELAWARE

ing, a perfectly simple square brick structure with its charmingly proportioned cupola and the balustraded deck, is as dignified and satisfying a public building as one can find remaining from the colonial period.

One of the principal residential streets is Orange Street, running along the western side of the Common. At Number 2 (pages 150 and 160), facing one end of the Court House, is the house now occupied by Dr. Booker but known historically as the Kensey Johns House, having been built for that gentleman in 1790. The façade is extremely simple but with a fenestration which gives great dignity. The low addition to the right adds considerable interest to the house; it was built as an office for Attorney Johns. This wing contains an entrance hall and one room, a room which with its delightful proportions and arched ceiling testifies to the taste of the builder. The main part of the house has the hall and stairs to the right, with a connecting door to the office, and on the left the living and dining rooms. The kitchen and service rooms are in a wing at the rear.

The paneling and mantels in the two principal rooms are simple, with rather Georgian character in the mould-

ings. The other walls of these rooms are plastered with a dado band carried around, vigorous mouldings on the door trims and paneled reveals at the windows. The stairs are interesting, particularly for the very simple manner in which the handrail forms a cap for the newel post.

On the façade we notice one peculiarity which is found in several other houses at New Castle; the marble lintels over the windows, though cut in profile to resemble a flat arch with a keystone, are really of one piece and without any cutting to imitate joints.

The owner is justly proud of the key plates and handles on the main doors, for it is maintained that they are the only ones of that particular design except those at Mount Vernon. In fact, a few years ago, he gave one of these to replace one that had disappeared from Mount Vernon.

Along this same street are several interesting brick façades. These are all similar in design, each has a simple, well-fenestrated wall, a sturdy, well-designed doorway, some richness at the cornice and usually a dormer with pilasters and a small pediment. With their interior paneling, even in rather simple, small rooms,

OLD DUTCH HOUSE—1665—NEW CASTLE, DELAWARE

VAN DYCK HOUSE, NUMBER 400 DELAWARE STREET, NEW CASTLE, DELAWARE

AMSTEL HOUSE—1730—NEW CASTLE, DELAWARE

Corner in Main Room
AMSTEL HOUSE, NEW CASTLE, DELAWARE

they show the refinement which went with the general affluence of the town during its heyday.

Also on Orange Street is a remainder of the earlier settlement, of a time nearer the pioneer days. It is a little house with its eaves not many feet above the sidewalk and is known as The Old Dutch House (page 145). Its date is not certain but it is presumed that it was built near the middle of the seventeenth century.

On the other side of the Common and one block beyond is the Strand, another street which was favored as the location for many of the better houses. This street runs more or less along the waterfront with the back yards of one side continuing down to the river in many places. It is on the Strand that we find the most pretentious house in New Castle, the Read House, which has been fully described in Volume IV, Chapter 9.

Near the Read House is the four-story building known as the Parish House (page 158). It was built for Charles Thomas a very few years after the completion

of the Read House. The façade facing the Strand is similar in detail to many others in the town, the pedimented doorway, the one-pieced keystoned lintels, the modillioned cornice and the single dormer. It does differ in that the doorway is in the middle instead of at one side where it would have permitted the maximum width for a principal room entered from the hallway. However, as it stands on a corner we find that the side elevation becomes quite important, though the designer seemed content with the balance of the front and made little attempt at an axial treatment on the side. The gable, spanning the broader dimension, is cut off at the top by a narrow deck or walk which was probably used for watching the shipping.

The detail of the doorway shows the mouldings and ornament which are almost identical with those of the Read House and were undoubtedly done by the same workmen. Few surfaces were left plain. Facias were treated with regular series of gouges, half-round

Living Room Mantel
VAN DYCK HOUSE, NEW CASTLE, DELAWARE

KENSEY JOHNS HOUSE, NEW CASTLE, DELAWARE

mouldings were carved with rope-like grooves and pitted with auger holes, while dentil courses were replaced by intricately devised bands of great richness.

In many ways the most interesting house in New Castle is Amstel House (pages 142 and 147) as the present owners, Prof. and Mrs. H. Hanby Hay, have named it, recalling one of the early names of the town. Very strongly Dutch in the feeling of its detail, it was built about 1730. The wide gable with an angle of about 29 degrees spans what is the main front, though now on a minor street. The rather heavy doorway, the wide muntins and the large curved frieze surface are more like various houses in and around Philadelphia than those in New Castle. There is a great deal of paneling on the interior and it too has a heaviness which is not found in the other houses, though at the same time it is quite interesting. (See pages 144 and 148).

The house at Number 400 Delaware Street (page 146), just opposite Amstel House, was built for Nicholas Van Dyck in 1799. It has in recent times been divided for two families by changing a window into a doorway as is seen in the illustration. The mantel (page 149) is quite ornate and in the piercing of the ornamental bands with auger holes takes its place with the work found in the Read House.

There are a number of other interesting houses along the few streets of New Castle which are similar to those shown here. Fortunately there seems a good chance that they will remain for some time to come, for not only is industry removed but in this older town there is a compactness, with the houses built as close to one another as in a city on plots of narrow frontage, which will keep the newcomer from squeezing in as he has on the larger plots of many of the New England towns.

There is, moreover, another reason for the probable permanence of the town as it now stands, in that there is a real pride and understanding in the community of the architectural heritage represented by these buildings, an appreciation of tradition which is in restful contrast to the incessant changes which are sweeping away so much of our colonial background. New Castle is still the complete setting for the simple and genteel life which brought these eighteenth-century houses into existence.

Paneling and Mantel in Dining Room
KENSEY JOHNS HOUSE, NEW CASTLE, DELAWARE

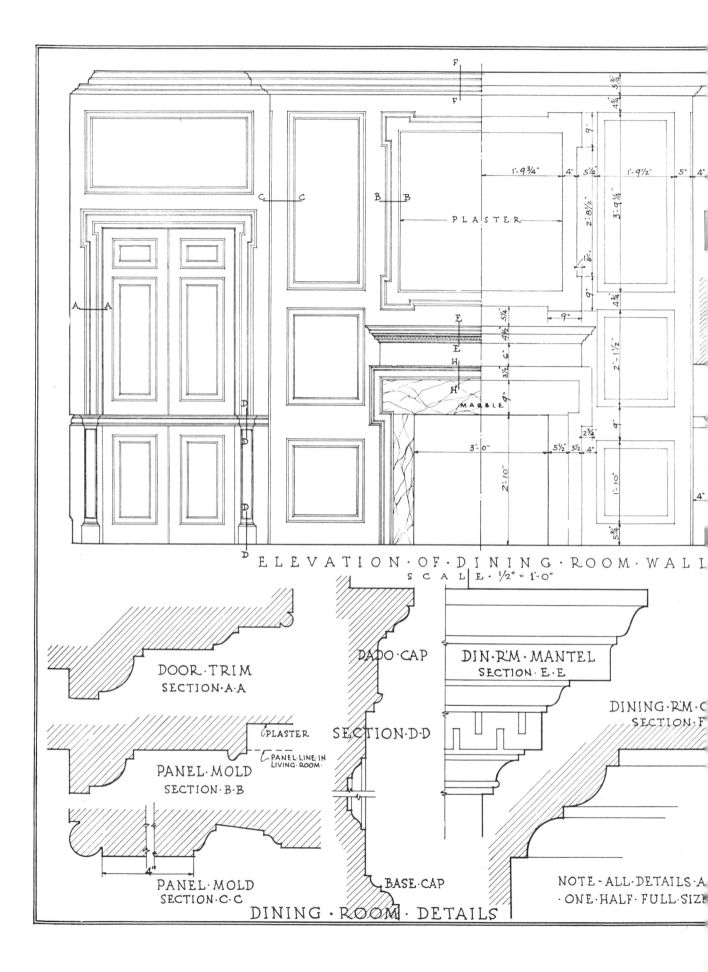

ELEVATION·OF·DINING·ROOM·WALL

SCALE·½"=1'-0"

PLASTER

MARBLE

DOOR·TRIM
SECTION·A·A

PANEL·MOLD
SECTION·B·B

PLASTER

PANEL·LINE·IN
LIVING·ROOM

PANEL·MOLD
SECTION·C·C

DADO·CAP

SECTION·D·D

BASE·CAP

DIN·R'M·MANTEL
SECTION·E·E

DINING·R'M·C
SECTION·F

NOTE-ALL·DETAILS·A
·ONE·HALF·FULL·SIZE

DINING·ROOM·DETAILS

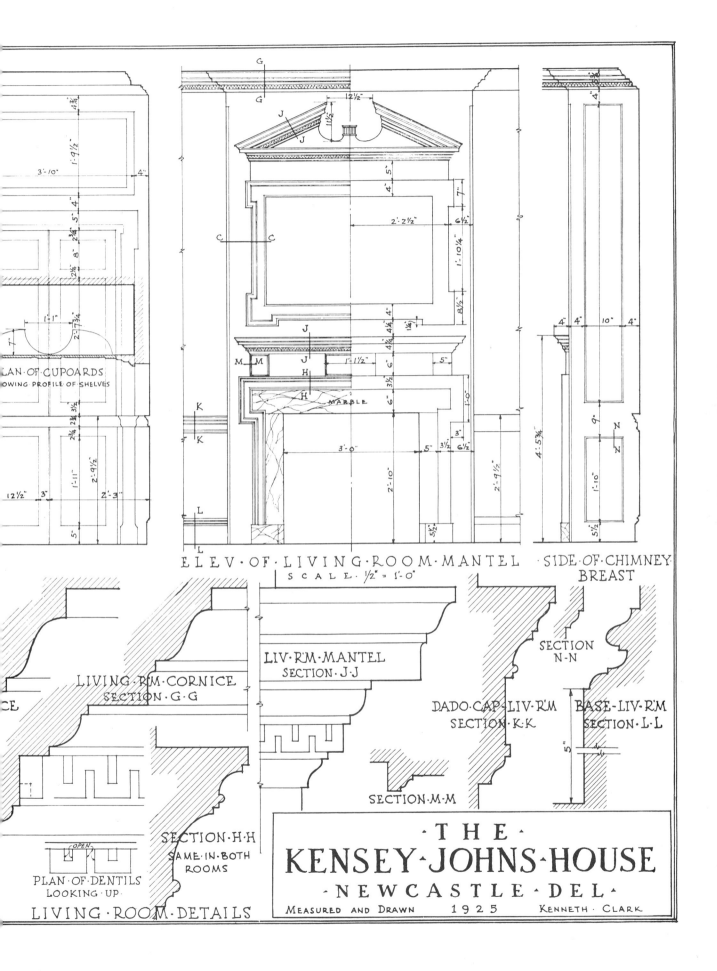

PLAN·OF·CUPOARDS
howing·PROFILE·OF·SHELVES

MARBLE

ELEV·OF·LIVING·ROOM·MANTEL
SCALE· ½" = 1'-0"

SIDE·OF·CHIMNEY·
BREAST

SECTION
N-N

LIVING·RM·CORNICE
SECTION·G·G

LIV·RM·MANTEL
SECTION·J·J

DADO·CAP·LIV·RM
SECTION·K·K

BASE·LIV·RM
SECTION·L·L

SECTION·M·M

SECTION·H·H
SAME·IN·BOTH
ROOMS

OPEN

PLAN·OF·DENTILS
LOOKING·UP

LIVING·ROOM·DETAILS

·THE·
KENSEY·JOHNS·HOUSE
·NEWCASTLE·DEL·
MEASURED AND DRAWN 1925 KENNETH·CLARK

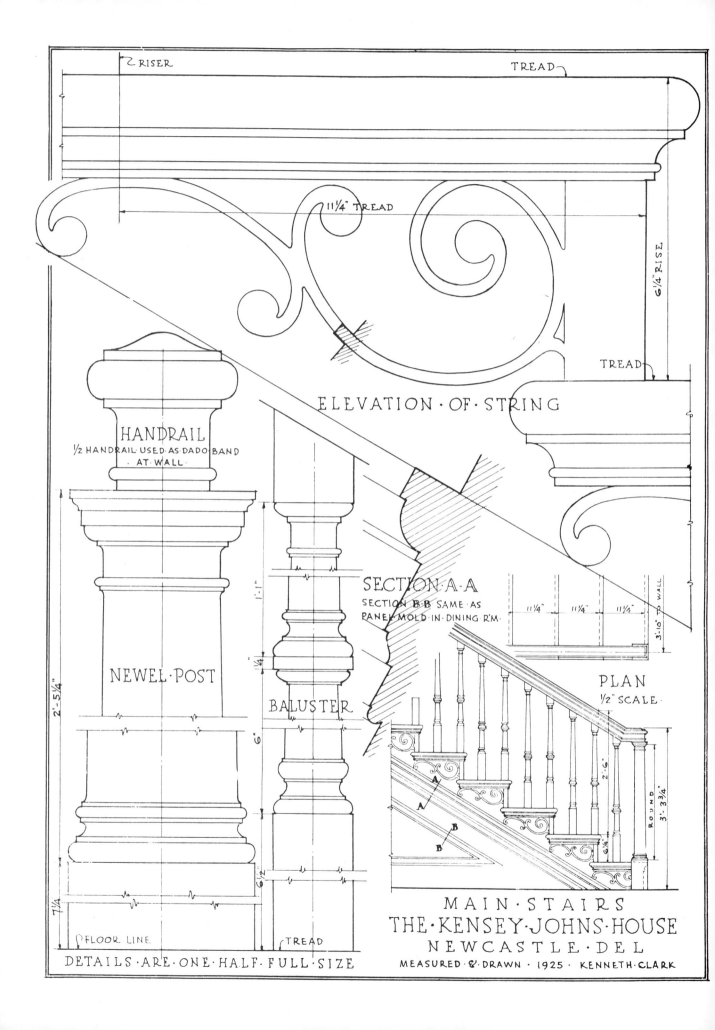

RISER

TREAD

11¼" TREAD

6¼" RISE

TREAD

ELEVATION · OF · STRING

HANDRAIL

½ HANDRAIL · USED · AS · DADO · BAND
· AT · WALL

SECTION · A·A
SECTION · B·B · SAME · AS
PANEL · MOLD · IN · DINING · R'M.

NEWEL · POST

BALUSTER

11¼" 11¼" 11¼"

3'-10" TO WALL

PLAN
½" SCALE ·

2'-5¼"

1'-1"

¼

6"

6½"

2'-6"

ROUND 3'-3¾"

6¼"

7¼

A

A

B

B

FLOOR LINE

TREAD

MAIN · STAIRS
THE · KENSEY · JOHNS · HOUSE
NEWCASTLE · DEL

DETAILS · ARE · ONE · HALF · FULL · SIZE

MEASURED · & · DRAWN · 1925 · KENNETH · CLARK

Living Room Mantel
KENSEY JOHNS HOUSE, NEW CASTLE, DELAWARE

Main Stairs

Detail of Living Room Mantel

KENSEY JOHNS HOUSE, NUMBER 2 ORANGE STREET, NEW CASTLE, DELAWARE

Window Detail, Kensey Johns House (1790)

Window Detail, Amstel House (1730)

WINDOW DETAILS – NEW CASTLE, DELAWARE

CHARLES THOMAS HOUSE — c1801 — NEW CASTLE, DELAWARE
Now called The Parish House

HOUSE AT NUMBER 18 ORANGE STREET, NEW CASTLE, DELAWARE

Detail of Entrance
KENSEY JOHNS HOUSE, NEW CASTLE, DELAWARE

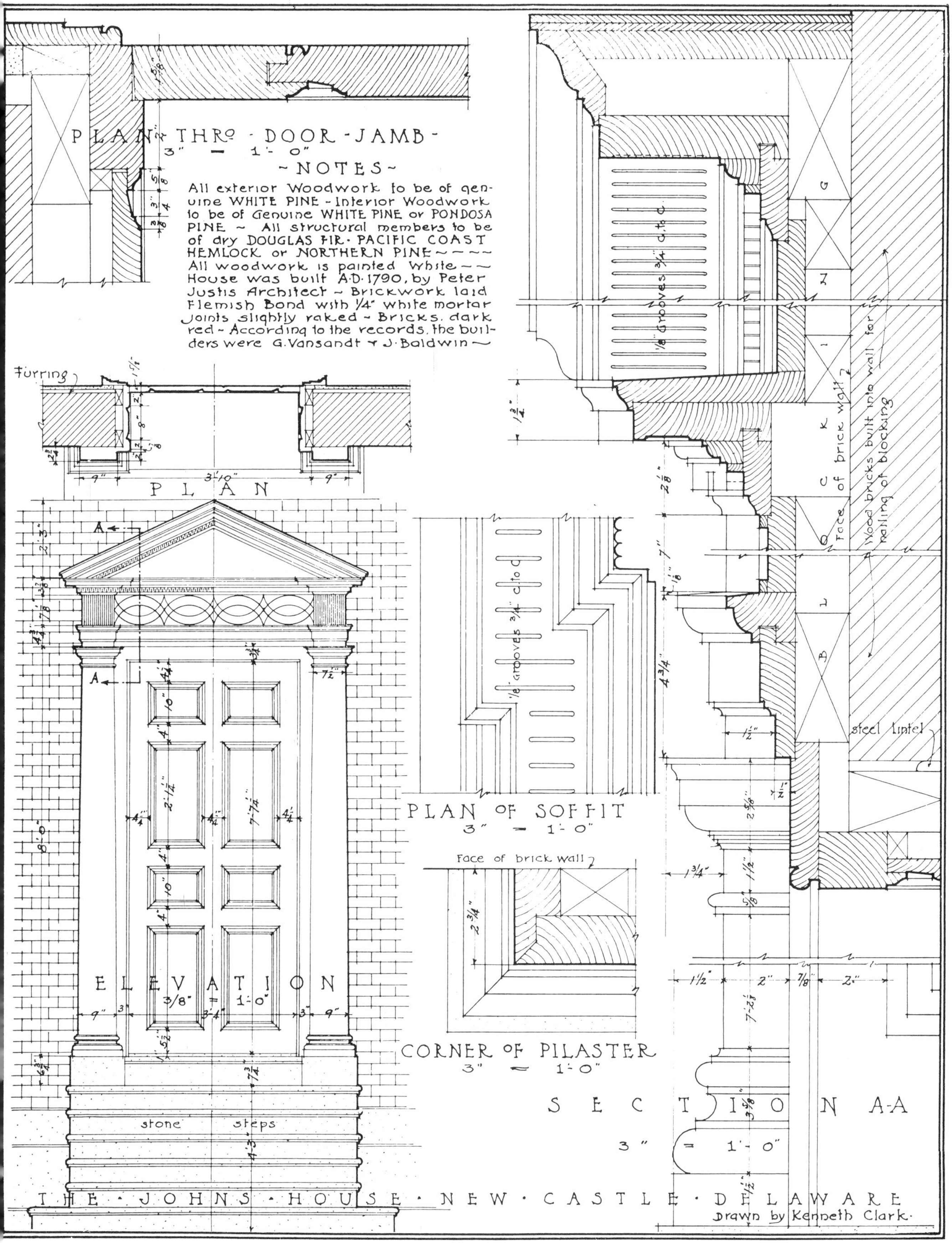

PLAN · THRO · DOOR · JAMB ·
3" = 1'-0"

~ NOTES ~

All exterior Woodwork to be of genuine WHITE PINE - Interior Woodwork
to be of Genuine WHITE PINE or PONDOSA
PINE ~ All structural members to be
of dry DOUGLAS FIR · PACIFIC COAST
HEMLOCK or NORTHERN PINE ~~~
All woodwork is painted White ~~
House was built A·D· 1790, by Peter
Justis Architect ~ Brickwork laid
Flemish Bond with 1/4" white mortar
joints slightly raked ~ Bricks, dark
red ~ According to the records, the builders were G.Vansandt + J.Baldwin ~

Furring

PLAN

A

A

ELEVATION
3/8" = 1'-0"

stone   steps

PLAN OF SOFFIT
3" = 1'-0"

Face of brick wall

CORNER OF PILASTER
3" = 1'-0"

1/8 grooves 3/4 c. to c.

Face of brick wall

Wood bricks built into wall for
nailing of blocking

steel lintel

SECTION A-A
3" = 1'-0"

THE · JOHNS · HOUSE · NEW · CASTLE · DELAWARE
Drawn by Kenneth Clark.

Cupola
TOWN HALL AND MARKET, NEW CASTLE, DELAWARE

George Read II House,
New Castle,
Delaware

Text by
Herbert C. Wise
Photographs by
Kenneth Clark
Originally published in 1925 as White Pine Monograph
Volume XI, Number 6

Detail of Front
GEORGE READ II HOUSE, NEW CASTLE, DELAWARE

THE GEORGE READ II HOUSE
AT NEW CASTLE, DELAWARE

WHEN the Swedish ships under the command of Peter Minuit anchored in the Delaware in 1638 their company of families bought from the Indians tracts which included the present sites of Wilmington and New Castle, and New Sweden was founded. The soil was fertile, game was abundant, and the settlement prospered. Even the arrival of Peter Stuyvesant in 1651 to claim the territory for the Dutch — erecting Fort Casimir by way of emphasis — and his naming the village New Amstel failed to shake the fortunes of the Swedes, for their individual holdings of land were not disturbed and they were allowed to continue their peaceful pursuits.

New Amstel was described as "a goodly town of about one hundred houses and containing a magazine, a guard house, a bake house and forge and residences for the clergymen and other officers." The settlement again fell into the hands of the Swedes and was again recovered by the Dutch; but upon Penn's arrival in 1682, he claimed the territory as a part of his Pennsylvania grant from the Duke of York, and thus terminated for all time the Swedish and Dutch authority upon the Delaware shores.

The town hall, the remains of the public market and, of course, the church and the court house can still be seen, as well as a number of delightful old residences. Of these may be mentioned the Amstel House, the Kensey Johns House, the Van Dyke House, the Church (or Thomas) House, but the largest and finest residence of the town is the Read House, pictures of which are to be found on these pages.

The house was commenced by George Read II in 1791 and completed in 1801. The mansion of his father, the first George Read, who was a signer of the Declaration of Independence, stood to the south (or left) of the present Read House in what is now the garden, and fronted on The Strand, as the street nearest the river is called. This house was destroyed by what the townsfolk call The Great Fire that swept New Castle in 1824 and destroyed some of the finest buildings.

The present mansion, erected by the son, occupies the northeast corner of a plot of ground having a frontage of about 180 feet on The Strand. The depth of the property is about 312 feet and extends to The Green. The walls are of brick, and it would be natural to suppose these were made at the southern end of the town long known as Brickmaker's Point — where roofing tiles were also made — if indeed the family records did not prove that they had been bought of Jeremiah Hornkett, brickmaker of Philadelphia, and transported down the river by shallop at the rate of one dollar per thousand. The bricks are of a uniform rich dark red like the traditional Philadelphia "stretchers." They measure 8½ × 2 × 4¼ inches which approaches the standard brick of this country rather than the shorter and higher bricks imported from England. Thomas Spikeman of Wilmington laid the bricks. James Traquair of Philadelphia was the stonecutter. The lumber was also bought in Philadelphia, and Peter Crowding of that city was the contractor for the carpenter work.

A plot of well kept lawn stretches before the house and gives a broad outlook on the River. This space also gives a satisfactory view of the stately façade, distinguished as it is by the simplicity of parts characteristic of the English Georgian style, the finely wrought wood detail of the entrance, the Palladian window and the main cornice, the dormers of a form that rings traditionally true, and above all the balustrade enclosing a "Captain's Walk." Examining the façade in detail, it is found to be 49 feet 2 inches in width. Granite steps and platform lead to the front entrance with its doorway 4 feet 5 inches wide. Together with sidelights and fluted pilasters of wood, the entrance measures 9 feet wide overall. An unusual device is to be seen in the divisions between the door and the sidelights. These are brought out to the face of the wall, thus recessing the door, as well as the sidelights, within deep paneled jambs and head of wood. The first story windows are 4 feet 3 inches wide; and as the bricks are so laid that five courses occupy 12¾ inches in height, the number of courses forming the jambs of the windows can be counted and the height of 8 feet 3½ inches ascertained. The marble window heads are really lintels with joint lines incised upon them to give the semblance of voussoirs. The ironwork appears to be of later date than that of the house itself.

The depth of the main or front body of the house is 46 feet 8 inches. A hallway 8 or 9 feet wide traverses the center of the house from the front entrance to the rear doorway opening upon the garden. The parlor and library divide equally the space upon the left of the hall. That upon the right is occupied by a square stair hall at the center, in front of which is the dining room and behind is the breakfast room. Beyond the last named and extending 50 feet or so further to the rear is the kitchen and service wing.

The interior doorways are provided with pilasters and entablatures in carved wood. With apologies to the the editor of a journal once devoted to white pine, we remark that the doors themselves are of mahogany. Their surrounding detail is of pine, however, and bears many coats of white paint. The design is derived from classic forms but here used with a freedom leading to an effect a modern architect might yearn in vain to realize, daring not to depart from his books. Should he do so his innovations would be adjudged unpardonable. Yet similar crudities are present here and criticism is stilled. The reason? Time has consecrated them. Then, too, the touch of the hand everywhere noticeable on these mouldings of long ago has laid upon them a pliancy and softness which dwell not in the products of modern planing mills and machinery.

In the frieze and centerpiece of a mantel "French putty" ornaments are found depicting a gentleman-at-arms being driven in a lion chariot, preceded by a flying messenger and followed by his armed retainers. Architraves, skirting and chair rails have generous proportions and heavy projections. The ornament and decoration of the woodwork continue through the first floor. In the library it is as elaborate as in the parlor. The dining room is simpler, with a mantel from which all the moulding decoration is omitted; but the stairway

LEGEND
A RECEPTION ROOM
B HALL
C DINING ROOM
D LIVING ROOM
E STAIR HALL
F LIBRARY
G PANTRY
H TERRACE
I KITCHEN
J WOOD HOUSE
K SMOKE HOUSE
L GARDEN

·FIRST·FLOOR·PLAN·

THE·GEORGE·READ·HOUSE
NEW·CASTLE·DELAWARE

and the second story hall are similar in treatment to the elaborate rooms of the first floor, yet they are quite distinct in detail. On the second floor, too, is the drawing room which is quite the most elaborate room of all and again displays distinct differences of detail. A fine ornamental frieze in moulded plaster is an added touch of decoration which distinguishes this room.

The hardware throughout the main part of the house is quite original, the escutcheons being formed of interlaced silver strands half-round in section. In one of the second floor bedrooms there is a quaint arrangement of wires running through pulleys permitting the brass bolt of the door to be opened by a person lying in bed.

Upon entering the house, if the weather be warm, one sees the garden beyond through a doorway elaborated with pilasters and semicircular transom with fanlight, and over all a horizontal cornice. Here is a brick-paved space under a grape arbor, and beyond them are greenhouse, potting house and tool house.

The garden was laid out in 1846 by Robert Buist. Its front portion, extending about 130 feet along The Strand and equaling the depth of the house, is laid out in three geometric parterres. Beyond this is the portion laid out in serpentine paths between cedar trees, box bushes, balsam firs and a great maple. Further on is the kitchen garden where at two corners stand English walnut trees. An aged Balm of Gilead, a magnolia macrophylla and a crepe myrtle are treasured landmarks.

In 1845 the property was bought by the Couper Family, of which a descendent, Miss Hettie Smith, was long the owner and occupant. Fortunately, the present owner, Mr. Philip Laird, is not only sensible of the architectural treasure in his keeping, but delights in its possession. Upon acquiring it he had it restored as far as possible to its original condition. This work was done under the intelligent and sympathetic direction of Brown and Whiteside, architects of Wilmington, who also added the brick wall surrounding the garden and the gateway.

GEORGE READ II HOUSE, NEW CASTLE, DELAWARE

Detail of Front Doorway
GEORGE READ II HOUSE, NEW CASTLE, DELAWARE

Detail of Palladian Window
GEORGE READ II HOUSE, NEW CASTLE, DELAWARE

Detail of Dormer
GEORGE READ II HOUSE, NEW CASTLE, DELAWARE

Hall
GEORGE READ II HOUSE, NEW CASTLE, DELAWARE

Detail of Hall
GEORGE READ II HOUSE, NEW CASTLE, DELAWARE

Mantel in Reception Room

GEORGE READ II HOUSE, NEW CASTLE, DELAWARE

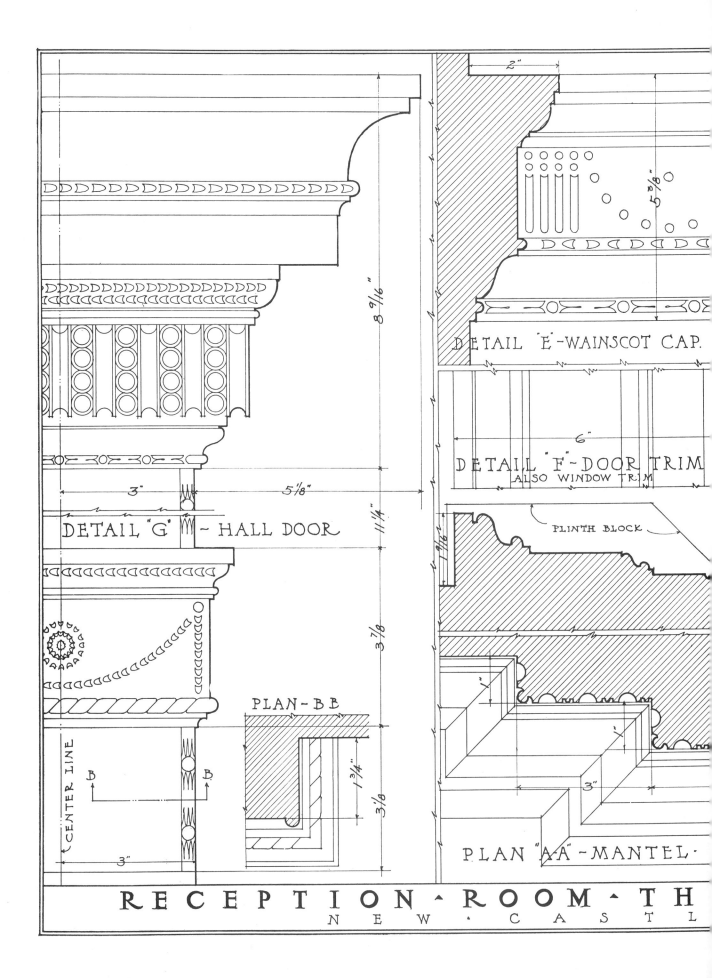

8 9/16"

DETAIL "G" — HALL DOOR

3"   5 1/8"

11 1/4"

3 7/8"

CENTER LINE

B   B

3"

PLAN — B B

1 3/4"

3/8"

2"

DETAIL "E" — WAINSCOT CAP.

5 3/8"

6"

DETAIL "F" — DOOR TRIM
ALSO WINDOW TRIM

9/16"

PLINTH BLOCK

1"

1"

3"

PLAN "A-A" — MANTEL.

RECEPTION · ROOM · TH
NEW · CASTL

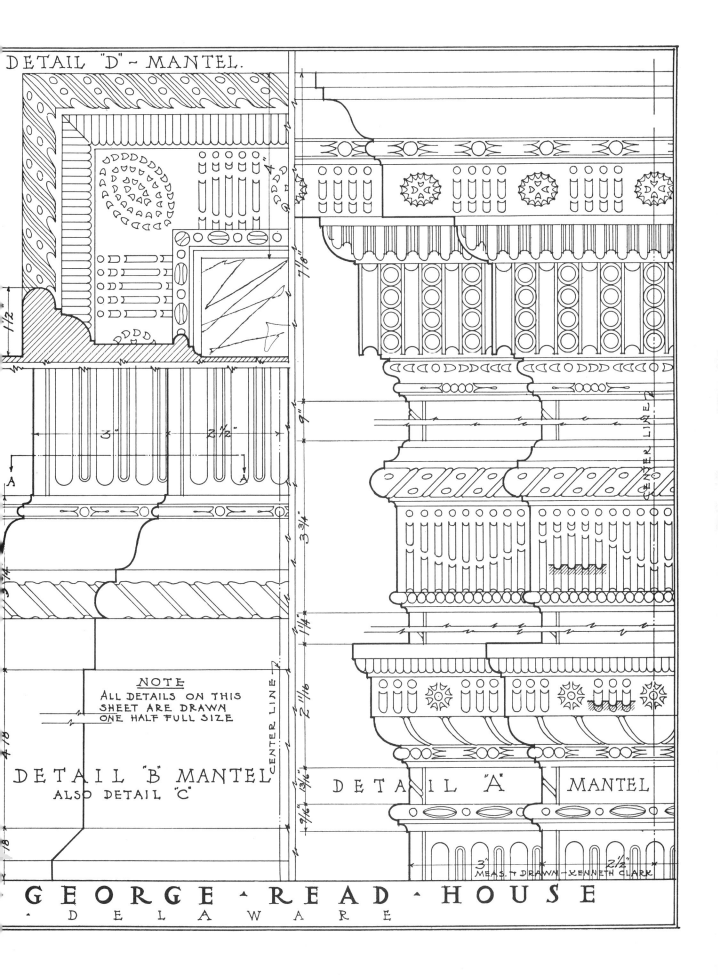

DETAIL "D" ~ MANTEL.

NOTE
All details on this
sheet are drawn
one half full size

DETAIL "B" MANTEL
also detail "C"

DETAIL "A" MANTEL

MEAS. + DRAWN + KENNETH CLARK

GEORGE · READ · HOUSE
· D E L A W A R E

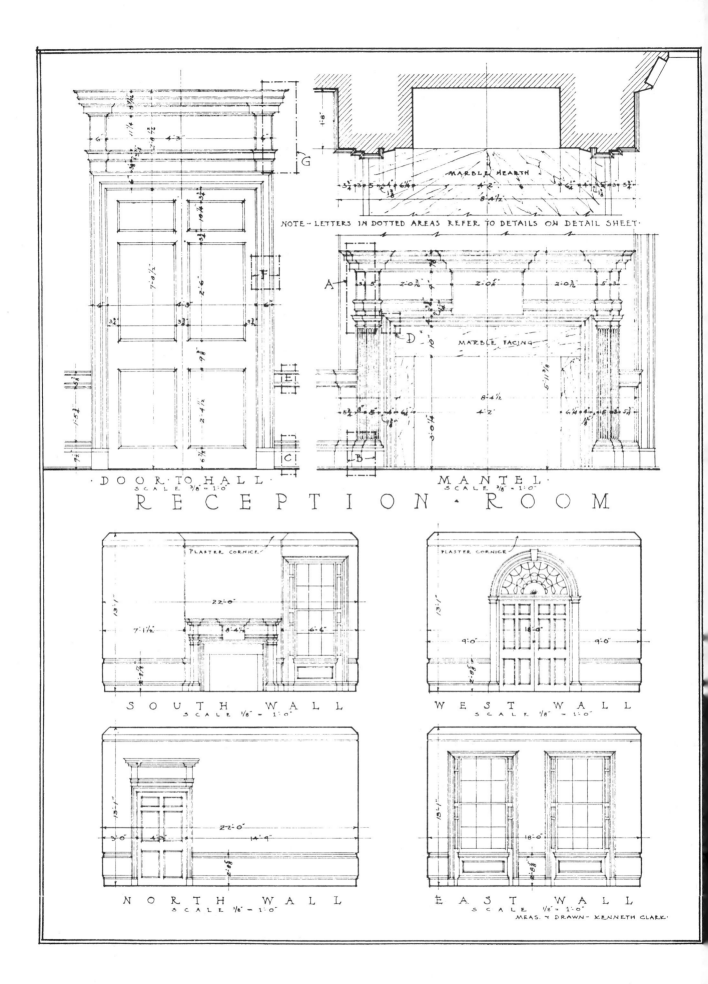

NOTE ~ LETTERS IN DOTTED AREAS REFER TO DETAILS ON DETAIL SHEET

MARBLE HEARTH

MARBLE FACING

·D O O R · T O · H A L L·
SCALE 3/8" = 1'0"

·M A N T E L·
SCALE 3/8" = 1'0"

R E C E P T I O N · R O O M

PLASTER CORNICE

·S O U T H   W A L L·
SCALE 1/8" = 1'0"

PLASTER CORNICE

·W E S T   W A L L·
SCALE 1/8" = 1'0"

·N O R T H   W A L L·
SCALE 1/8" = 1'0"

·E A S T   W A L L·
SCALE 1/8" = 1'0"

MEAS. & DRAWN ~ KENNETH CLARK

Detail A Mantel
GEORGE READ II HOUSE, NEW CASTLE, DELAWARE

Detail of Doorway Between Reception and Living Rooms
GEORGE READ II HOUSE, NEW CASTLE, DELAWARE

Detail of Stairway at First Floor Level
GEORGE READ II HOUSE, NEW CASTLE, DELAWARE

Detail of Second Floor Drawing Room
GEORGE READ II HOUSE, NEW CASTLE, DELAWARE

Detail of Doorway, Second Story Hall
GEORGE READ II HOUSE, NEW CASTLE, DELAWARE

Interior Doorway
GEORGE READ II HOUSE, NEW CASTLE, DELAWARE

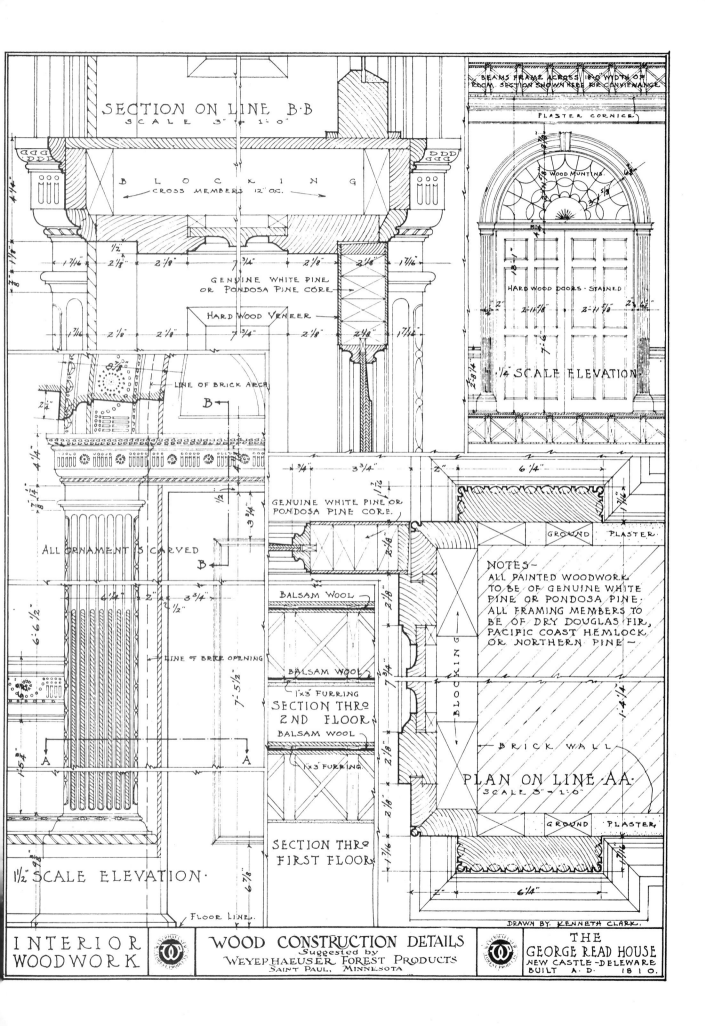

SECTION ON LINE B·B
SCALE 3" = 1'-0"

BLOCKING
CROSS MEMBERS 12" O.C.

GENUINE WHITE PINE
OR PONDOSA PINE CORE

HARD WOOD VENEER

LINE OF BRICK ARCH

B

ALL ORNAMENT'S CARVED

B

LINE OF BRICK OPENING

A   A

1½" SCALE ELEVATION·

FLOOR LINE.

BEAMS FRAME ACROSS 18'-0" WIDTH OF
ROOM. SECTION SHOWN HERE FOR CONVENIENCE

PLASTER CORNICE

WOOD MUNTINS

HARD WOOD DOORS - STAINED

¼ SCALE ELEVATION

GENUINE WHITE PINE OR
PONDOSA PINE CORE.

BALSAM WOOL

BALSAM WOOL

1x3 FURRING
SECTION THRO
2ND FLOOR
BALSAM WOOL

1x3 FURRING

SECTION THRO
FIRST FLOOR

GROUND   PLASTER

NOTES-
ALL PAINTED WOODWORK
TO BE OF GENUINE WHITE
PINE OR PONDOSA PINE.
ALL FRAMING MEMBERS TO
BE OF DRY DOUGLAS FIR,
PACIFIC COAST HEMLOCK
OR NORTHERN PINE -

BLOCKING

BRICK WALL

PLAN ON LINE A·A·
SCALE 3" = 1'-0"

GROUND   PLASTER

DRAWN BY. KENNETH CLARK

INTERIOR
WOODWORK

WOOD CONSTRUCTION DETAILS
Suggested by
WEYERHAEUSER FOREST PRODUCTS
SAINT PAUL, MINNESOTA

THE
GEORGE READ HOUSE
NEW CASTLE - DELEWARE
BUILT A.D. 1810.

Garden Entrance Detail
GEORGE READ II HOUSE, NEW CASTLE, DELAWARE

THE
BUILDER's COMPANION,

AND

WORKMAN's GENERAL ASSISTANT:

DEMONSTRATING,

After the moſt eaſy and practical Method,

ALL THE

PRINCIPAL RULES of ARCHITECTURE,

FROM

The PLAN to the ORNAMENTAL FINISH;

Illuſtrated with a greater Number of uſeful and familiar Examples than **any** Work of that Kind hitherto publiſhed;

WITH

Clear and ample INSTRUCTIONS annexed to each Subject or Number on the ſame PLATE;

Being not only uſeful but neceſſary to all MASONS, BRICKLAYERS, PLASTERERS, CARPENTERS, JOINERS, and others concerned in the ſeveral Branches of BUILDING, &c.

The Whole correctly engraven on 102 Folio COPPER-PLATES,

Containing upwards of SEVEN HUNDRED DESIGNS on the following Subjects, &c.

I. Of Foundations, Walls, and their Diminutions, Fitneſs of Chimneys, and Proportion of Light to Rooms, with the due Scantlings of Timber to be cut for Building, &c.

II. Great Variety of Geometrical, Elliptic, and Polygon Figures, with Rules for their Formation. Centering of all Sorts for Groinds, Brick and Stone Arches, &c. both Circular and Splay'd, alſo with Circular Sofits in a Circular Wall: Many Examples for Glewing and Vaneering, Niches, &c. with Rules for tracing the Cover of Curve-Line Roofs, Piers, Vaſes, Pedeſtals for Sun-dials, Buſts, &c. and their moſt ſuitable Proportions.

III. General Directions for Framing Floors and Partitions, Truſs-roofs, &c. and Methods to find the Length and Backing of Hips, ſtrait or curve Lines to any Pitch, Square or Bevel.

IV. Of Stair-Caſes, variouſly conſtructed; the Methods of working Ramp and Twiſt-Rails—Profils of Stairs to ſhew the Manner of ſetting Carriages for the Steps, alſo the framing of String-Boards and Rails, and likewiſe of fixing them.

V. The Five Orders of Architecture from *Palladio*, with the Rule for gauging Flutes and Fillets on a diminiſh'd Column, by a Method extremely eaſy, and intirely new.

VI. Doors, Windows, Frontiſpieces, Chimney-pieces, Cornices, Mouldings, &c. truly proportion'd, in a plain and gen-eel taſte.

VII. Sacred Ornaments, *viz.* Altar-pieces, Pulpits, Monuments, &c.

VIII. Gothic Architecture, being a various Collection of Columns, Entablatures, Arches, Doors Windows, Chimney-Pieces, and other Decorations in that prevailing Taſte—And it may be noted of theſe, as of all the foregoing Examples, that they are immediately adapted to Workmen, and may be executed by the meaneſt Capacity.

By *WILLIAM PAIN*, Architect and Joiner.

LONDON:

Printed for the AUTHOR, and ROBERT SAYER, at the Golden Buck in *Fleet-Street*,

MDCCLXII.

Facsimile of Original Title Page

FOREWORD

A VITAL IMPETUS was given to architecture in the eighteenth century by the number of British architectural hand-books which appeared. The accepted manner in England was soon echoed in America and the master builders and designers here manifestly looked to the English pattern books, published and republished since 1700 for their designs of moldings, cornices, entablatures; for portals and even for facades.

A careful examination of these early works would seem to indicate that they were not such complete guides as to leave no need for creative ability on the part of the individual who used them. They were, with rare exceptions, merely books of "the orders" and were not manuals of English architecture.

A part of the working library of one of America's justly famed carpenter-builders of the eighteenth century has recently been discovered. The oldest of these volumes is dated 1724. Many of the treatises present a similarity of plan. They first offer elementary problems in geometry after which are included plates of the five orders and in addition details of construction and such architectural elements as windows, doors and mantels. By the omission of suggestions for plan arrangement, and, in most books, the complete facade design, it was probably implied that the responsibility for the design of an ensemble rested with the individual to whose capacity these hand-books were a "remembrancer."

The knowledge that these books on architecture were owned by "practising architects" in America in the middle of the eighteenth century strengthens our conviction that the hand-books were generally within arm's reach of the amateur designer. They were published at a time when almost every man of culture in England interested himself in architecture and when a high standard of lay criticism existed.

Believing that those interested in the sources of colonial work would enjoy having a reprint of one of the best and least familiar of these books, we have selected *"The Builder's Companion"* by William Pain, which was published in London in 1762 and which was in use in the Colonies in 1763.

A majority of the folio copper plates in this rare work will be published, in this and the next number of THE MONOGRAPH SERIES. The two parts will furnish the reader with a document which should be interesting not only to the lovers of Colonial architecture but also as a working tool in the drafting room. The first twelve plates are selected from the illustrations of the Five Orders. These are followed by Frontispieces, Proportions of Doors and Windows. The second part will contain the Pediments, Piers for Gates, Vases, Sun Dials, the proportions of Cornices to Rooms, Mouldings, Mantels, Stair Cases, Pulpits, Etc.

The modern designer will be convinced, we feel sure, that William Pain, Architect and Joiner, endeavored to catch the spirit of classic proportion and "by an intire New Scale" to show the significance of the orders and to make it easy for anyone to adapt the proportions to modern usage. Our early architecture was not molded by buildings of the mother country so much as by the engraved specimens of the Italian Renaissance and Roman orders. To study the following Plates, not so much with a view to achieving their exact reproduction, but rather to acquire the sense of elegance, proportion and delicacy with which so many of them are redolent, will more than repay the student of architecture.

THE EDITOR

NOTE: This foreward appeared in the 1931 reprint of William Pain's *The Builder's Companion*, published as White Pine Monograph Volume XVII, Number 1. We have included Russell Whitehead's foreward because it provides background on early builder's handbooks in general, and William Pain's work in particular.

Cornices for Rooms or Eves of Houses or any place Required.

25 Cornices for Rooms or Eves of Houses with the Grose Measures, all figured; how many parts Each Cornice is to be Divided into, and them parts disposed to each Member, as they are figured in height and projection. To find the Grose measures of the Cornices in height divide from the Floor to the Ceiling into 15 parts, take one for the Cornice, or divide into 16 parts, one is the Cornice; or if the height be divided into 35 parts take two for the Cornice and in some Cases the height may be divided into 18 parts then take one for the Cornice. any of these Grose Measures may be used at pleasure according to the places they are Required to set on the projections, drop a plumb Line as a, b, then set Back 2½ and 4½ then drop one at c, set Back 2¾ and so for all the rest which is plain to Inspection.

REPRODUCTION of Copper-plate engraving by WILLIAM PAIN, Architect and Joiner
PUBLISHED IN LONDON, 1762

PREFACE

READER.

 S in all Things Order is to be obferved, fo efpecially in this excellent Art of Architecture it is requifite that every Part and Member have its right Order and due Proportion . There have been may Mafters who with great Care and Induftry have brought this Art to a great Perfection, among whom the famous PALLADIO deferves to be placed in the higheft Rank by all judicious Artifts ; therefore for the Benefit of Workmen, and that it may be made more ufeful for all Artificers in Building and Lovers of this moft noble Art, I have laid down the five Orders of Architecture according to PALLADIO by an intire NEW SCALE, to proportion the Orders to any given Height, and to find the Model or Diameter of the Column, with every Part of the Orders by the faid Scale ; and for the better underftanding of which, the Reader is defired to take notice, that by the Model is fignified the Meafure of the whole Diameter of the Column : As for Example ; Let the Diameter be twelve, fifteen, or eighteen Inches for the Model, to be divided into fixty Parts or Minutes, as may be feen by the Scale on the Diameter of the Column in Page 32, that Scale meafuring every Member in the Order, which will be proportionable one to another; this dividing of the Diameter into fixty Parts or Minutes, muft be ufed in all the Orders, in which I have, with my utmoft Endeavours, rendered it very intelligible, with a great Number of other ufeful Things for the Benefit of Workmen.

The Tuscan Order

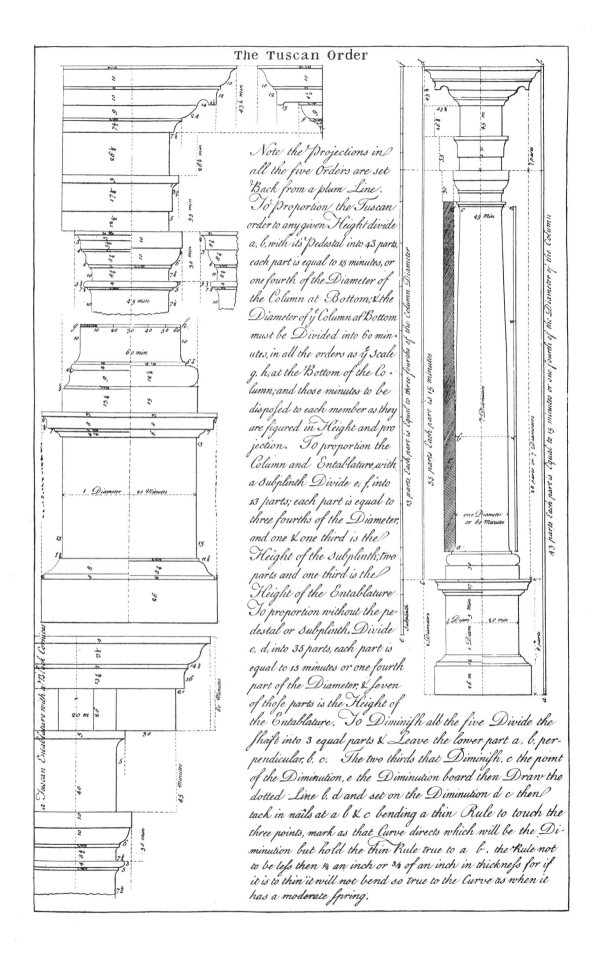

Note the Projections in all the five Orders are set Back from a plum Line. To Proportion the Tuscan order to any given Height divide a, b, with its Pedestal into 43 parts, each part is equal to 15 minutes, or one fourth of the Diameter of the Column at Bottom; & the Diameter of ÿ Column at Bottom must be Divided into 60 minutes, in all the orders as ÿ Scale g, h, at the Bottom of the Column; and those minutes to be disposed to each member as they are figured in Height and projection. To proportion the Column and Entablature, with a Subplinth Divide e, f, into 13 parts; each part is equal to three fourths of the Diameter, and one & one third is the Height of the subplinth; two parts and one third is the Height of the Entablature. To proportion without the pedestal or subplinth, Divide c, d, into 35 parts, each part is equal to 15 minutes or one fourth part of the Diameter, & seven of those parts is the Height of the Entablature. To Diminish all the five Divide the shaft into 3 equal parts & Leave the lower part a, b, perpendicular, b, c. The two thirds that Diminish, c the point of the Diminution, e the Diminution board then Draw the dotted Line b, d and set on the Diminution d c then tack in nails at a b & c bending a thin Rule to touch the three points, mark as that Curve directs which will be the Diminution but hold the thin Rule true to a b. the Rule not to be less then ½ an inch or ¾ of an inch in thickness for if it is to thin it will not bend so true to the Curve as when it has a moderate spring.

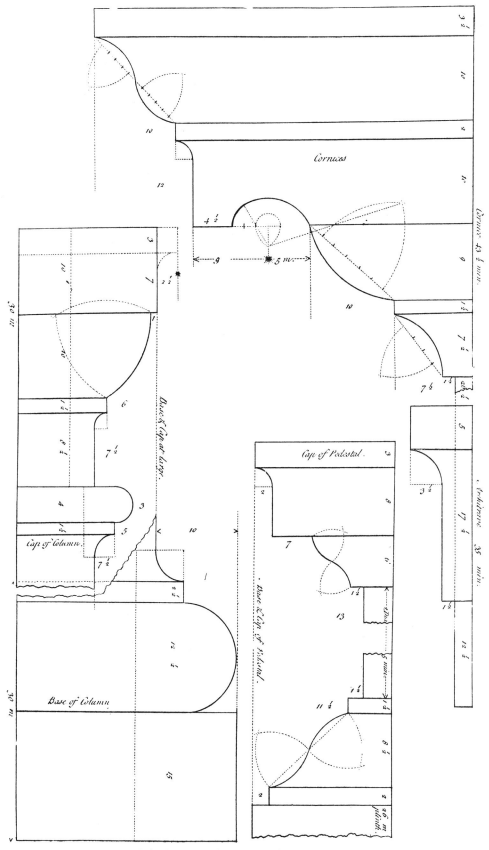

Mouldings at large for the Tuscan Order

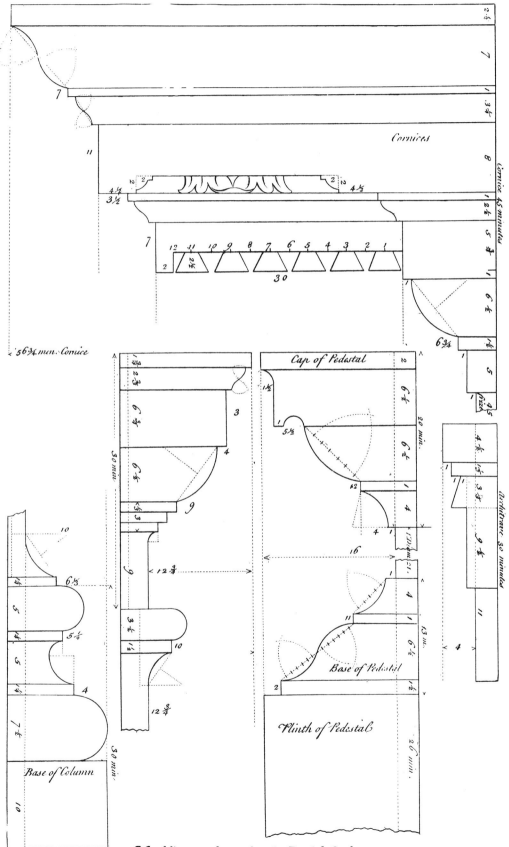

Mouldings at large for the Dorick Order

The Dorick Order.

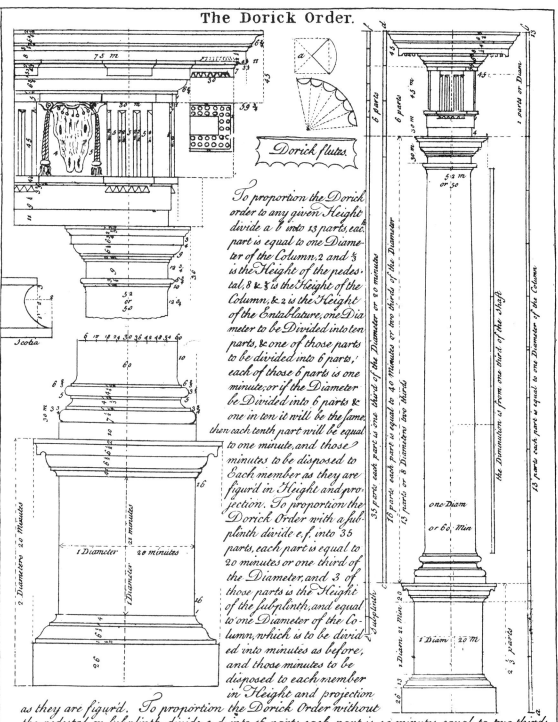

Dorick flutes.

Scotia

To proportion the Dorick order to any given Height divide a b into 13 parts, each part is equal to one Diameter of the Column, 2 and ⅔ is the Height of the pedestal, 8 & ⅔ is the Height of the Column, & 2 is the Height of the Entablature, one Diameter to be Divided into ten parts, & one of those parts to be divided into 6 parts; each of those 6 parts is one minute; or if the Diameter be Divided into 6 parts & one in ten it will be the same, then each tenth part will be equal to one minute, and those minutes to be disposed to Each member as they are figur'd in Height and projection. To proportion the Dorick Order with a fubplinth divide e, f, into 35 parts, each part is equal to 20 minutes or one third of the Diameter, and 3 of those parts is the Height of the fubplinth, and equal to one Diameter of the Column, which is to be divided into minutes as before, and those minutes to be disposed to each member in Height and projection as they are figur'd. To proportion the Dorick Order without the pedestal or fubplinth, divide c d into 16 parts, each part is 40 minutes, equal to two thirds of the Diameter, but if the Column is but 8 Diameters high and set on a pedestal then divide a b into 36 parts each part is 20 minutes, if 8 Diameters, when set on a fubplinth, divide e f into 11 parts, each part is one Diameter, without pedestal or fubplinth if 8 Diameters divide c d into ten parts each part is one Diameter, figure a is the manner of Striking the Dorick flutes as the Ancients did make them without fillets

The Ionick Order.

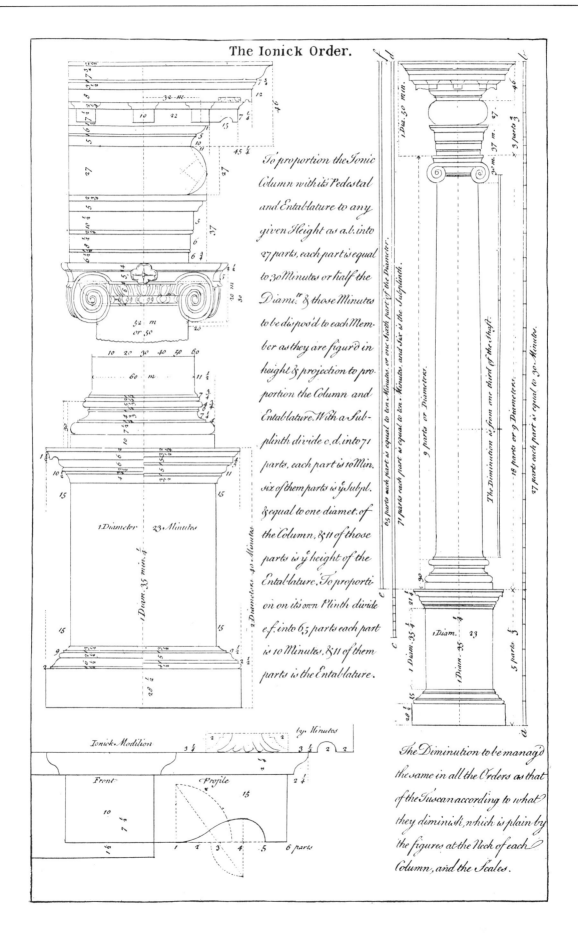

To proportion the Ionic Column with its Pedestal and Entablature to any given Height as a.b. into 27 parts, each part is equal to 30 Minutes or half the Diamr. & those Minutes to be dispos'd to each Member as they are figur'd in height & projection to proportion the Column and Entablature. With a Subplinth divide c.d. into 71 parts, each part is 10 Min. six of them parts is ÿ Subpl. & equal to one diamet. of the Column, & 11 of those parts is ÿ height of the Entablature. To proportion on its own Plinth divide e.f. into 65 parts each part is 10 Minutes, & 11 of them parts is the Entablature.

The Diminution to be manag'd the same in all the Orders as that of the Tuscan according to what they diminish, which is plain by the figures at the Neck of each Column, and the Scales.

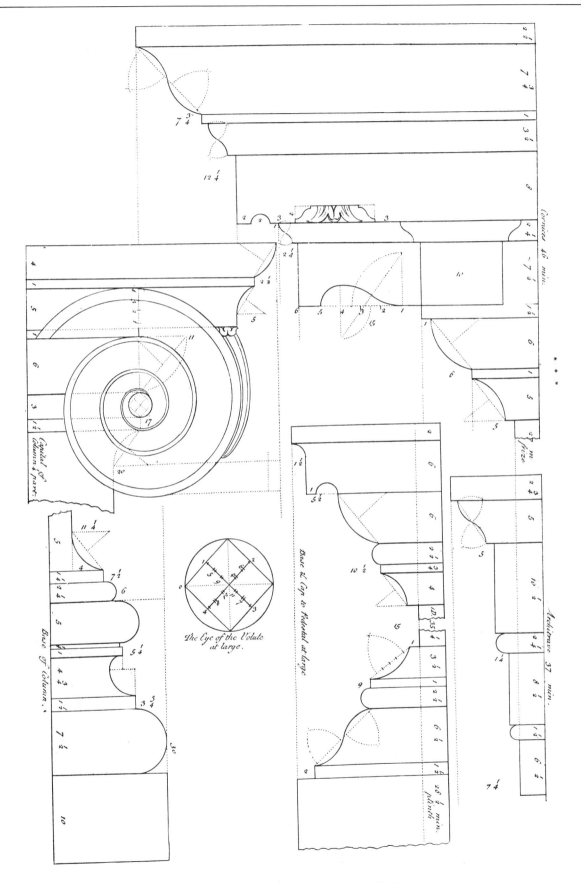

Mouldings at large for the Ionick Order

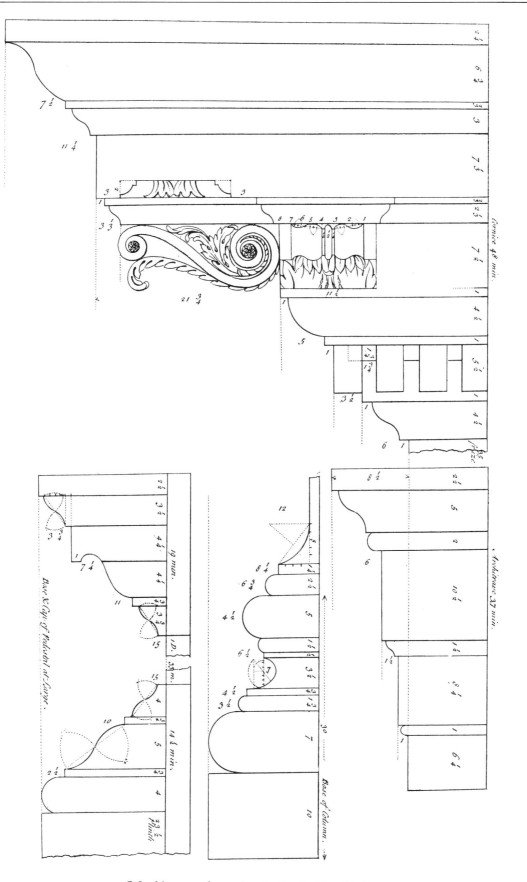

Mouldings at large for the Corinthian Order

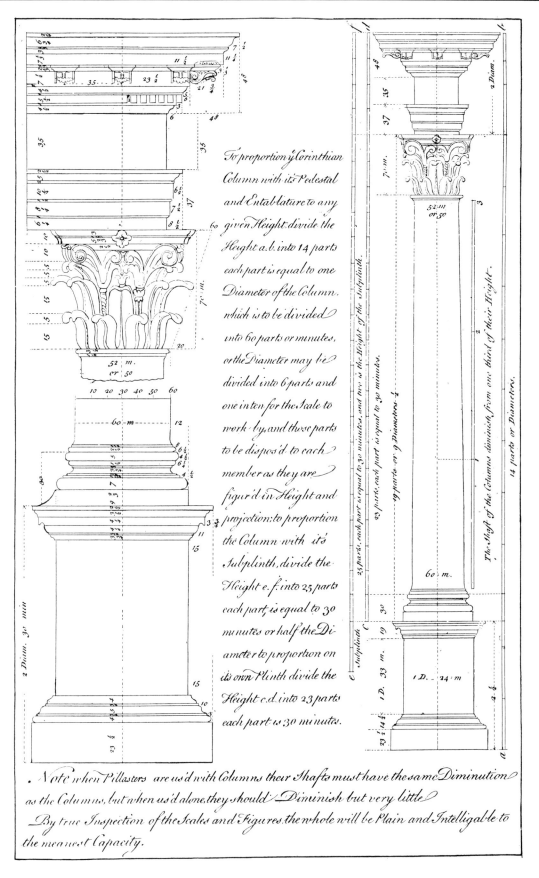

To proportion ÿ Corinthian Column with it's Pedestal and Entablature to any given Height divide the Height a.b. into 14 parts each part is equal to one Diameter of the Column, which is to be divided into 60 parts or minutes, or the Diameter may be divided into 6 parts and one in ten for the Scale to work by, and those parts to be dispos'd to each member as they are figur'd in Height and projection: to proportion the Column with it's Subplinth, divide the Height e. f. into 25 parts each part is equal to 30 minutes or half the Diameter to proportion on its own Plinth divide the Height c.d. into 23 parts each part is 30 minutes.

. Note when Pillasters are us'd with Columns their Shafts must have the same Diminution as the Columns, but when us'd alone, they should Diminish but very little

By true Inspection of the Scales and Figures, the whole will be Plain and Intelligable to the meanest Capacity.

The Corinthian Order

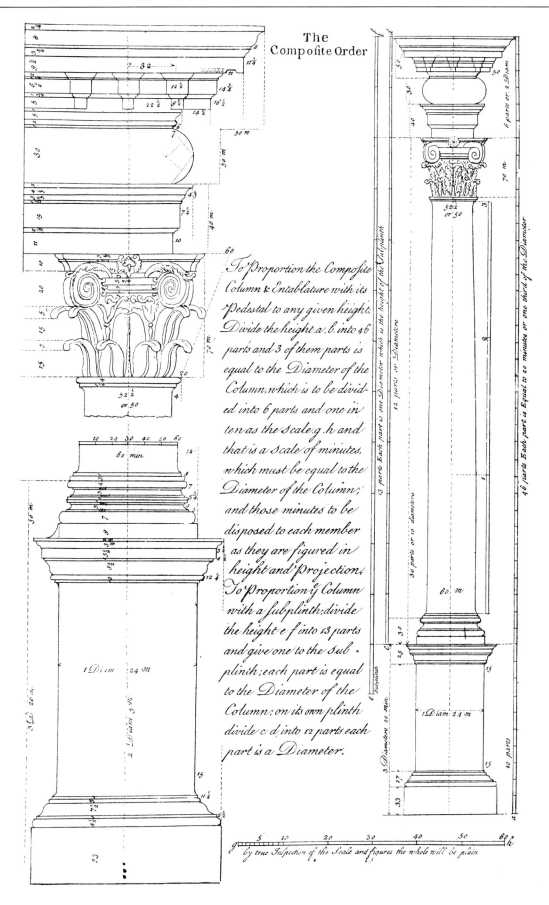

The
Composite Order

To Proportion the Composite Column & Entablature with its Pedestal to any given height. Divide the height a.b. into 46 parts and 3 of them parts is equal to the Diameter of the Column, which is to be divided into 6 parts and one in ten as the scale g.h and that is a scale of minutes, which must be equal to the Diameter of the Column; and those minutes to be disposed to each member as they are figured in height and Projection. To Proportion y Column with a subplinth; divide the height e.f into 13 parts and give one to the subplinth; each part is equal to the Diameter of the Column; on its own plinth divide c.d into 12 parts each part is a Diameter.

by true Inspection of the Scale and figures the whole will be plain

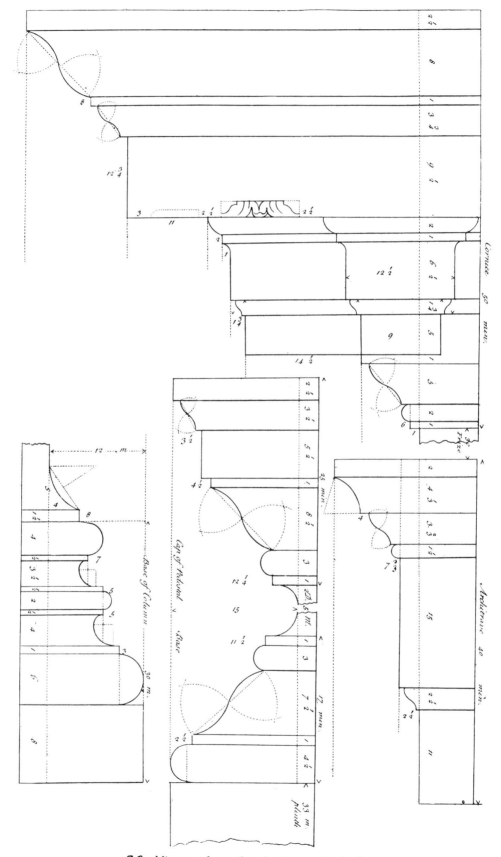

Mouldings at large for the Composite Order

The Ionick volute & the Bases of the Columns at Large with the Centers
To strike each molding. The Eye of the volute at large with the Centers
shewn & this * is the Centers for the inside Line of the list, which is one
fifth part between the Centers, when pillasters are fluted, divide the
breadth into 29 parts give 3 to a flute & 1 to a fillet but if beads at the Angle,
divide in 31 parts &c.

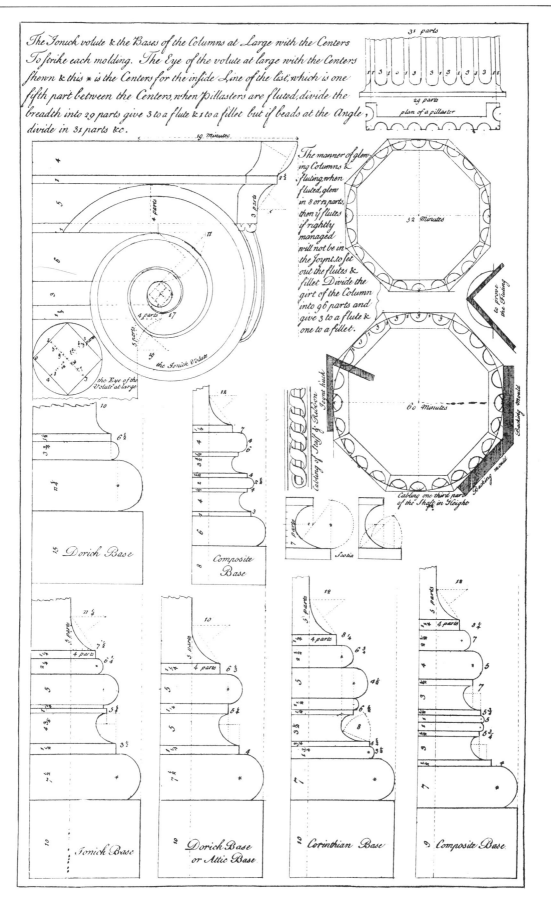

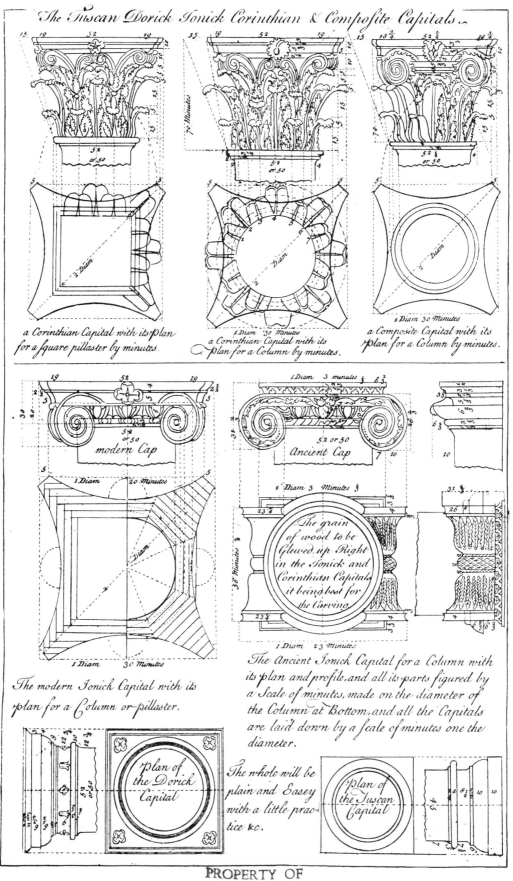

The Tuscan Dorick Ionick Corinthian & Composite Capitals.

a Corinthian Capital with its plan for a square pillaster by minutes.

a Corinthian Capital with its plan for a Column by minutes.

a Composite Capital with its plan for a Column by minutes.

modern Cap

Ancient Cap

The modern Ionick Capital with its plan for a Column or pillaster.

The grain of wood to be Glewed up Right in the Ionick and Corinthian Capitals it being best for the Carving

The Ancient Ionick Capital for a Column with its plan and profile, and all its parts figured by a Scale of minutes, made on the diameter of the Column at Bottom, and all the Capitals are laid down by a scale of minutes one the diameter.

The whole will be plain and Easey with a little practice &c.

Plan of the Dorick Capital

Plan of the Tuscan Capital

PROPERTY OF
HIGH POINT PUBLIC LIBRARY
HIGH POINT, NORTH CAROLINA

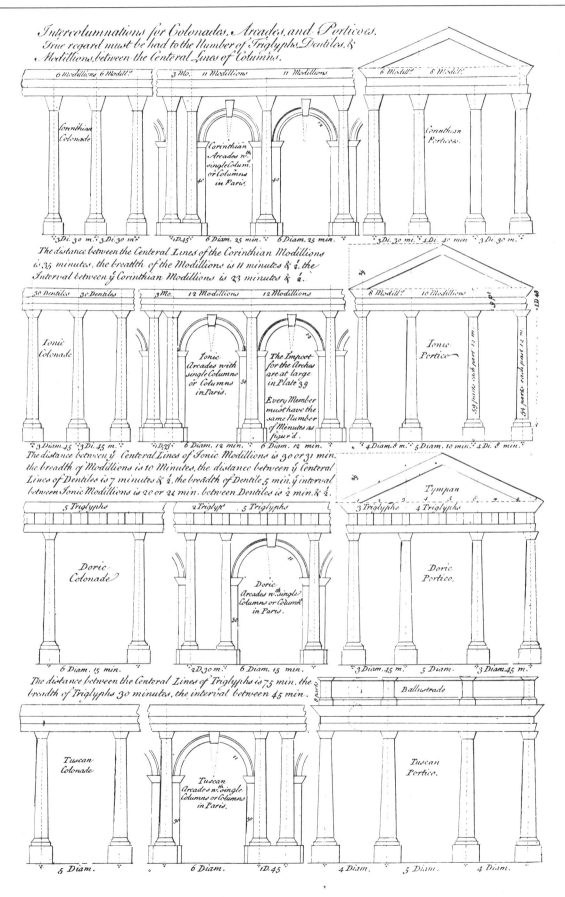

Intercolumnations for Colonades, Arcades, and Porticoes.
True regard must be had to the Number of Triglyphs, Dentiles, &
Modillions, between the Central Lines of Columns.

6 Modillions 6 Modill.   3 Mo. 11 Modillions 11 Modillions   6 Modill. 8 Modill.

Corinthian
Colonade

Corinthian
Arcades n.th
single Colum.ns
or Columns
in Paris.

Corinthian
Porticoes.

3 Di. 30 m. 3 Di. 30 m.   1 D. 45 6 Diam. 25 min. 6 Diam. 25 min.   3 Di. 30 mi. 4 Di. 40 min 3 Di. 30 m.

The distance between the Central Lines of the Corinthian Modillions
is 35 minutes, the breadth of the Modillions is 11 minutes & ½, the
Interval between ÿ Corinthian Modillions is 23 minutes & ½.

30 Dentiles 30 Dentiles   3 Mo. 12 Modillions 12 Modillions   8 Modill. 10 Modillions

Ionic
Colonade

Ionic
Arcades with
single Columns
or Columns
in Paris.

The Impost
for the Arches
are at large
in Plate 39

Every Member
must have the
same Number
of Minutes as
figur'd.

Ionic
Portico

3 Diam. 45 3 Di. 45 m.   1 D. 35 6 Diam. 12 min. 6 Diam. 12 min.   4 Diam. 8 m. 5 Diam. 10 min. 4 Di. 8 min.

The distance between ÿ Central Lines of Ionic Modillions is 30 or 31 min.
the breadth of Modillions is 10 Minutes, the distance between ÿ Central
Lines of Dentiles is 7 minutes & ½, the breadth of Dentile 5 min. ÿ interval
between Ionic Modillions is 20 or 21 min. between Dentiles is 2 min. & ½.

5 Triglyphs   2 Triglyphs 5 Triglyphs   Tympan
3 Triglyphs 4 Triglyphs

Doric
Colonade

Doric
Arcades n.th single
Columns or Column.s
in Paris.

Doric
Portico.

6 Diam. 15 min.   2 D. 30 m. 6 Diam. 15 min.   3 Diam. 45 m. 5 Diam. 3 Diam. 45 m.

The distance between the Central Lines of Triglyphs is 75 min. the
breadth of Triglyphs 30 minutes, the interval between 45 min.

Ballustrade

Tuscan
Colonade

Tuscan
Arcades n.th single
Columns or Columns
in Paris.

Tuscan
Portico.

5 Diam.   6 Diam. 1 D. 45   4 Diam. 5 Diam. 4 Diam.

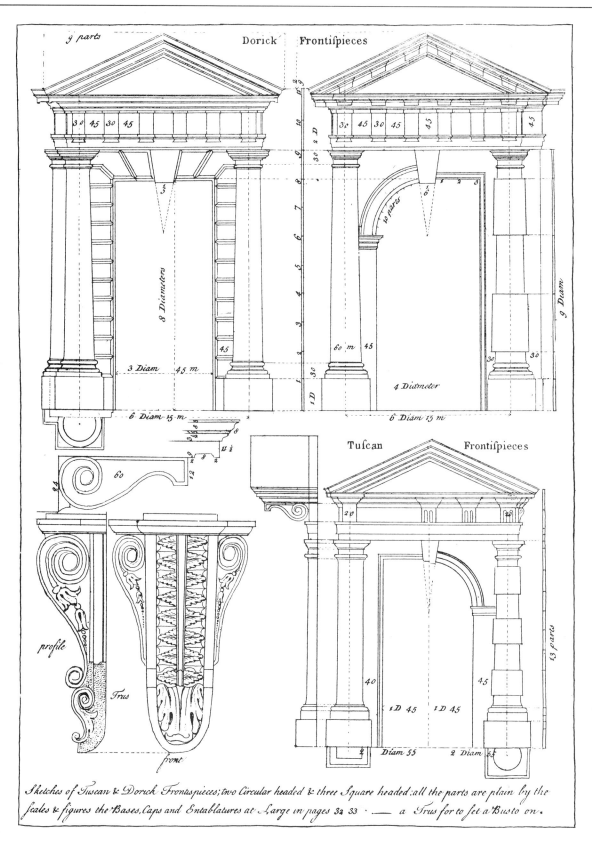

Sketches of Tuscan & Dorick Frontispieces; two Circular headed & three Square headed: all the parts are plain by the scales & figures the Bases, Caps and Entablatures at Large in pages 32 33 · —— a Trus for to set a Busto on.

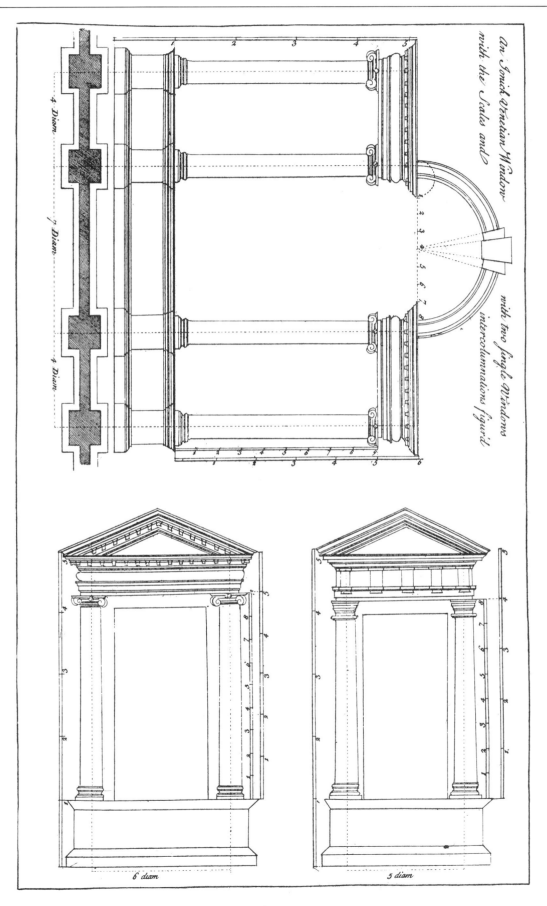

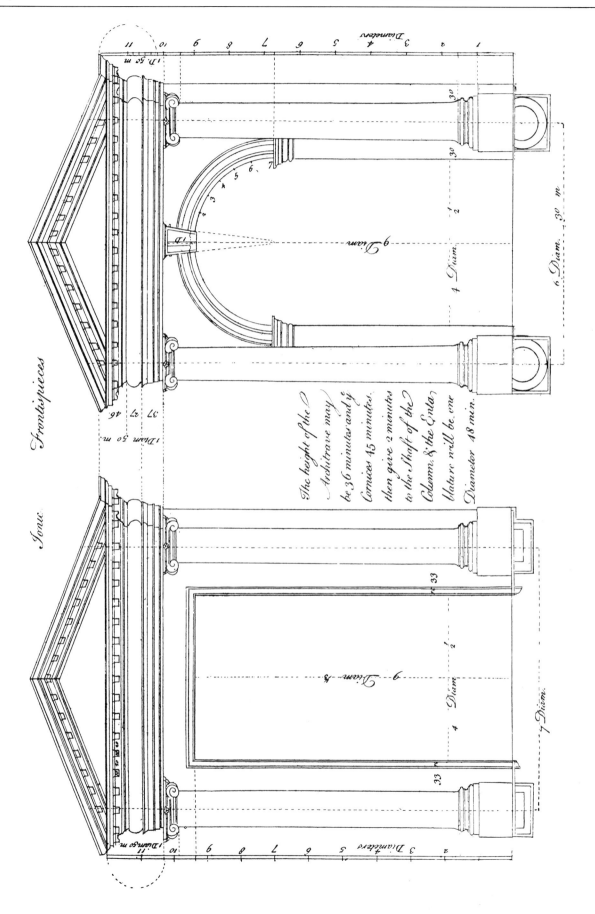

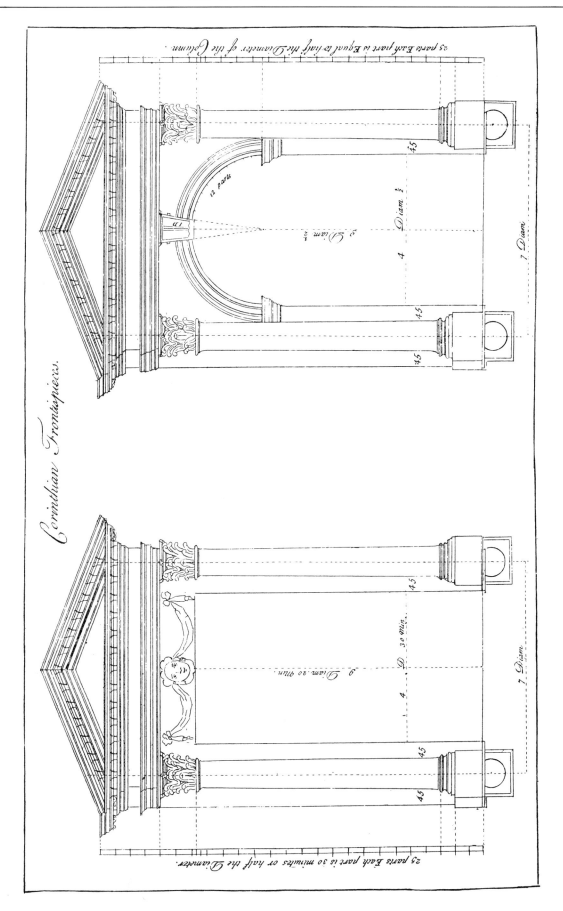

Corinthian Frontispieces.

The proportion
of Doors

To proportion Frontispieces to Doors. That mark'd A divides the opening or width of the door into six parts, give one to the Architrave, then divide the Architrave into three parts, give one to the knee of the Architrave and two to the open pillaster, then for the height of the Architrave, frize and Cornice which is half the width of the opening of the door, this is plain by Circles, the frize and Cornices take their parts from the Architrave. Suppose the Scale a . b to be the width of the Architrave G. which must allways be divided into Six parts, then give five of them parts to the frize and Seven to the Cornices as they are figured, or five and a half to the frize and six and a half to the Cornices as that of B. The height of the opening is two diameters one Sixth or two diameters one twelfth and the Rest will be plain to inspection &c . The length of the knee, twice the Architrave breadth. the profile of the truss ⅗ of the Architraves breadth or equal to the breadth of the Architrave.

The Proportion of Doors & Windows

The manner of Striking Raking Cornices as at B. divide the height of the Raking mold o.4 into four equal parts, and draw the lines parallel with the Raking, then make the projections all equal as a b. then take the Projections from the Level Cornices as a.b.c.d.e.f.g.h; set them on the Raking molding as a b.4 1.3 2. 2 5.1 6 that will give the Curve of the molding, and for upper Return'd molding, Set on the same parts square from the Back, to meet the Raking Lines, and that will be the Curve of the Return'd molding, then at those points tack in Nails, and Bend a thin slip to the Nails, and mark it by, that will be the Curve Line or faces of the moldings g.r hear is three doors and two windows which is plain by the parts and figures.——— The Length of the knee twice the Breadth of the Archatrave for y margents of doors divide the width of the door into 9 parts and give one to the margent and the Bottom Raile and Lock Raile double the margent.

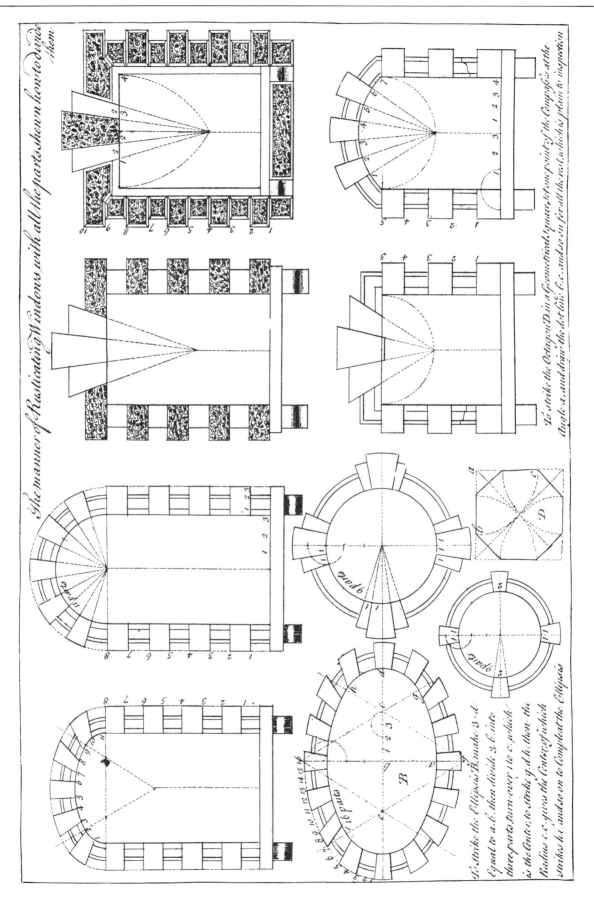

THE
BUILDER's COMPANION,

AND

WORKMAN's GENERAL ASSISTANT:

DEMONSTRATING,

After the moſt eaſy and practical Method,

ALL THE

PRINCIPAL RULES of ARCHITECTURE,

FROM

The PLAN to the ORNAMENTAL FINISH;

Illuſtrated with a greater Number of uſeful and familiar Examples than any Work of that Kind hitherto publiſhed;

WITH

Clear and ample INSTRUCTIONS annexed to each Subject or Number on the ſame PLATE;

Being not only uſeful but neceſſary to all MASONS, BRICKLAYERS, PLASTERERS, CARPENTERS, JOINERS, and others concerned in the ſeveral Branches of BUILDING, &c.

The Whole correctly engraven on 102 Folio COPPER-PLATES,

Containing upwards of SEVEN HUNDRED DESIGNS on the following Subjects, &c.

I. Of Foundations, Walls, and their Diminutions, Fitneſs of Chimneys, and Proportion of Light to Rooms, with the due Scantlings of Timber to be cut for Building, &c.

II. Great Variety of Geometrical, Elliptic, and Polygon Figures, with Rules for their Formation. Centering of all Sorts for Grounds, Brick and Stone Arches, &c. both Circular and Splay'd, alſo with Circular Sofits in a Circular Wall: Many Examples for Glewing and Vaneering, Niches, &c. with Rules for tracing the Cover of Curve-Line Roofs, Piers, Vaſes, Pedeſtals for Sun-dials, Buſts, &c. and their moſt ſuitable Proportions.

III. General Directions for Framing Floors and Partitions, Truſs-roofs, &c. and Methods to find the Length and Backing of Hips, ſtrait or curve Lines to any Pitch, Square or Bevel.

IV. Of Stair-Caſes, variouſly conſtructed; the Methods of working Ramp and Twiſt-Rails—Profils of Stairs to ſhew the Manner of ſetting Carriages for the Steps, alſo the framing of String-Boards and Rails, and likewiſe of fixing them.

V. The Five Orders of Architecture from Palladio, with the Rule for gauging Flutes and Fillets on a diminiſh'd Column, by a Method extremely eaſy, and intirely new.

VI. Doors, Windows, Frontiſpieces, Chimney-pieces, Cornices, Mouldings, &c. truly proportion'd, in a plain and genteel taſte.

VII. Sacred Ornaments, viz. Altar-pieces, Pulpits, Monuments, &c.

VIII. Gothic Architecture, being a various Collection of Columns, Entablatures, Arches, Doors Windows, Chimney-Pieces, and other Decorations in that prevailing Taſte—And it may be noted of theſe, as of all the foregoing Examples, that they are immediately adapted to Workmen, and may be executed by the meaneſt Capacity.

By *WILLIAM PAIN*, Architect and Joiner.

LONDON:

Printed for the AUTHOR, and ROBERT SAYER, at the Golden Buck in *Fleet-Street*,
MDCCLXII.

The manner of open Pediments with Busto's & Shells for the open part of the PEDIMENT.

Piers for Gates.

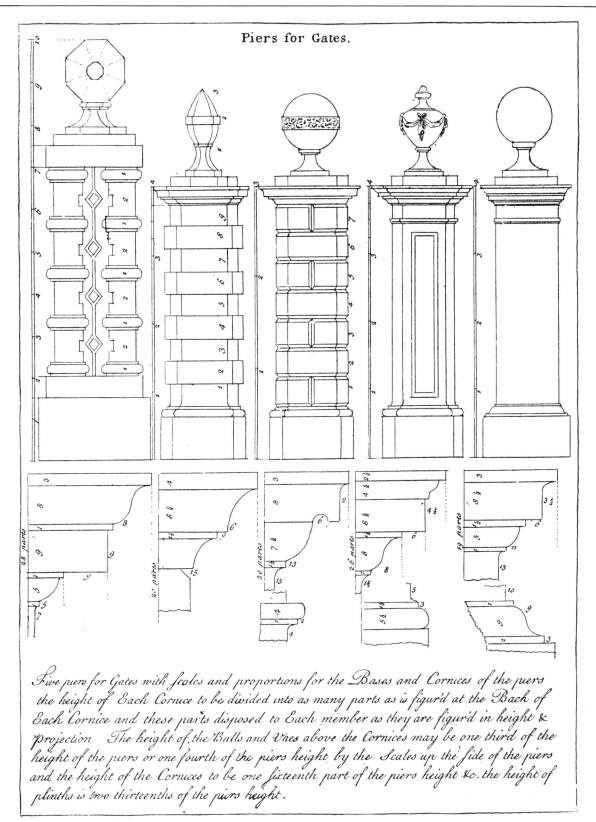

Five piers for Gates with scales and proportions for the Bases and Cornices of the piers
the height of Each Cornice to be divided into as many parts as is figur'd at the Back of
Each Cornice and these parts disposed to Each member as they are figur'd in height &
projection The height of the Balls and Vaes above the Cornices may be one third of the
height of the piers or one fourth of the piers height by the Scales up the side of the piers
and the height of the Cornices to be one sixteenth part of the piers height &c. the height of
plinths is two thirteenths of the piers height.

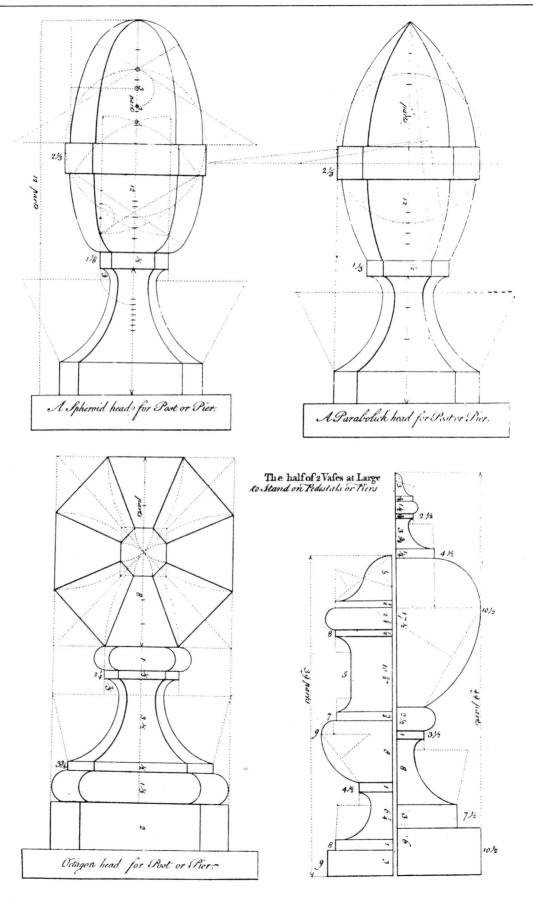

A Spheroid head for Post or Pier.

A Parabolick head for Post or Pier.

The half of 2 Vases at Large
to Stand on Pedestals or Piers

Octagon head for Post or Pier.

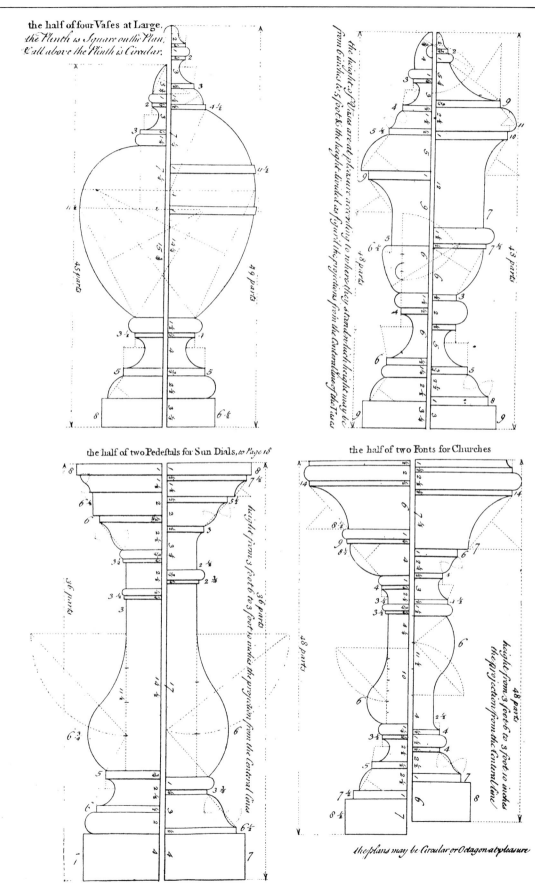

the half of four Vases at Large.
the Plinth is Square on the Plan,
& all above the Plinth is Circular.

the half of two Pedestals for Sun Dials, *to Page 18*

the half of two Fonts for Churches

the plans may be Circular or Octagon at pleasure

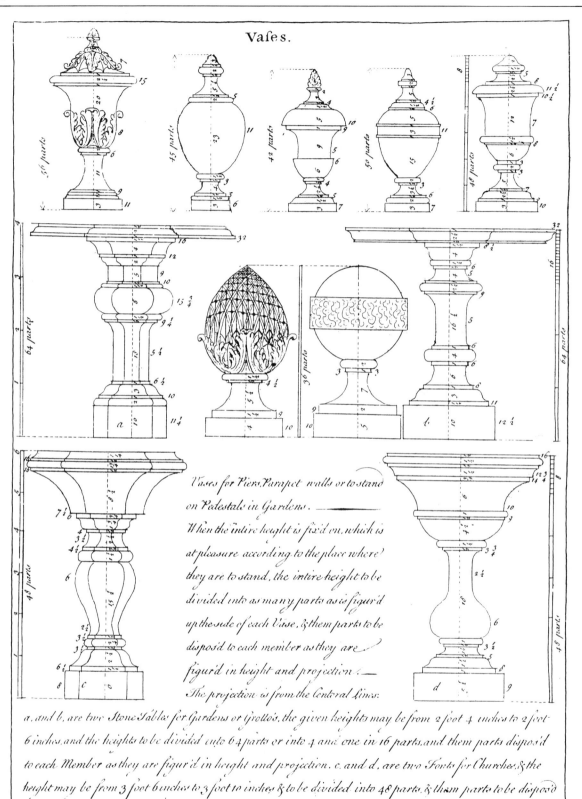

Vases.

Vases for Piers, Parapet walls or to stand
on Pedestals in Gardens. ———
When the intire height is fix'd on, which is
at pleasure according to the place where
they are to stand, the intire height to be
divided into as many parts as is figur'd
up the side of each Vase, & them parts to be
dispos'd to each member as they are
figur'd in height and projection ——
The projection is from the Central Lines.

a, and b, are two Stone Tables for Gardens or Grotto's, the given heights may be from 2 foot 4 inches to 2 foot
6 inches, and the heights to be divided into 64 parts or into 4 and one in 16 parts, and them parts dispos'd
to each Member as they are figur'd in height and projection. c. and d, are two Fonts for Churches, & the
height may be from 3 foot 6 inches to 3 foot 10 inches & to be divided into 48 parts, & them parts to be dispos'd
to each Member as they are figur'd in height & projection, the projection is from the Central Lines, &c. ———

Four Pedestals for Sun Dials. The heights may be from three foot ten Inches to four foot, and that height to be divided into six parts, one of which parts is to be divided into twelve parts, and those parts to be disposed to each member as they are figured in height and projection. The projections are all set from the Central Lines, a. and b. is two pedestals for Bustos: Proportion by the same Ruls the heights from four to five foot, and the parts to be disposed to each member as they are figur'd in height and projection.

Pedestals for Sun-Dials and Bustos

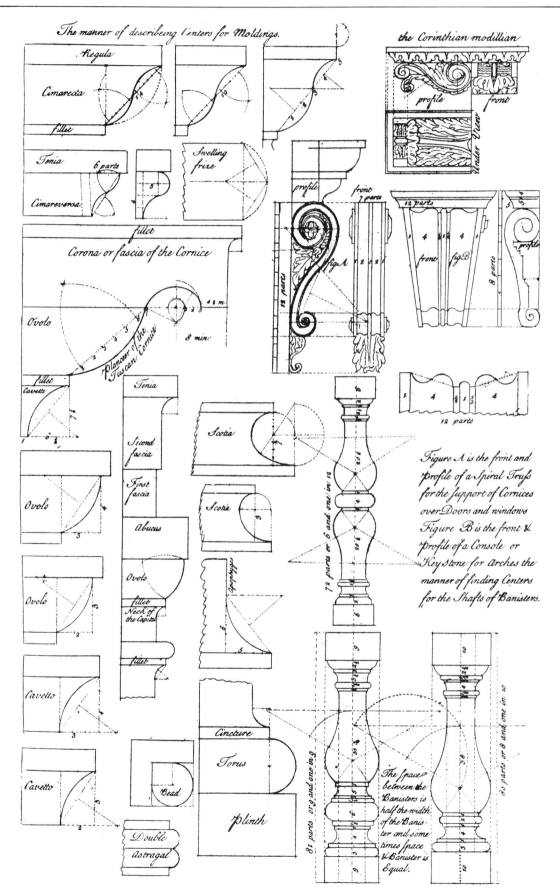

The manner of describeing Centers for Moldings.

the Corinthian modillian

Figure A is the front and profile of a Spiral Truss for the support of Cornices over Doors and windows Figure B is the front & profile of a Console or Key stone for Arches the manner of finding Centers for the Shafts of Banisters.

The space between the Banisters is half the width of the Banister and some times space & Banister is Equal.

The proportion of Cornices & margents to Rooms.

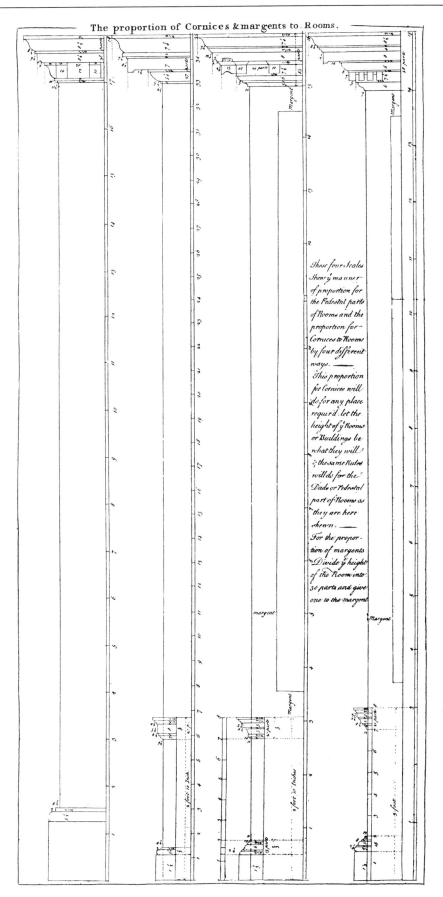

These four Scales
Shew ye manner
of proportion for
the Pedestal parts
of Rooms and the
proportion for
Cornices to Rooms
by four different
ways.

This proportion
for Cornices will
do for any place
requir'd, let the
height of ye Rooms
or Buildings be
what they will
; the same Rules
will do for the
Dado or Pedestal
part of Rooms as
they are here
shewn.

For the propor-
tion of margents
Divide ye height
of the Room into
30 parts and give
one to the margent

Moldings for Doors Windows and Chimneys.

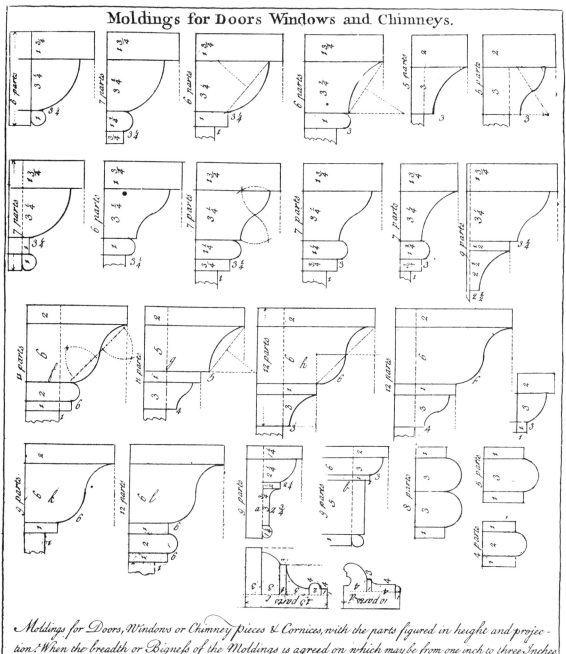

Moldings for Doors, Windows or Chimney pieces & Cornices, with the parts figured in height and projection: When the breadth or Bigness of the Moldings is agreed on which may be from one inch to three Inches, and them marked a.b.c.d. the breadths may be from three Inches to six Inches, and the Cornices marked f.g.h.i.k.l. the breadths or heights, may be from two Inches to six Inches: the breadth or height to be divided into as many parts as is figured at the Back of Each molding, & dispose the same parts to the Projections of Each molding &c.

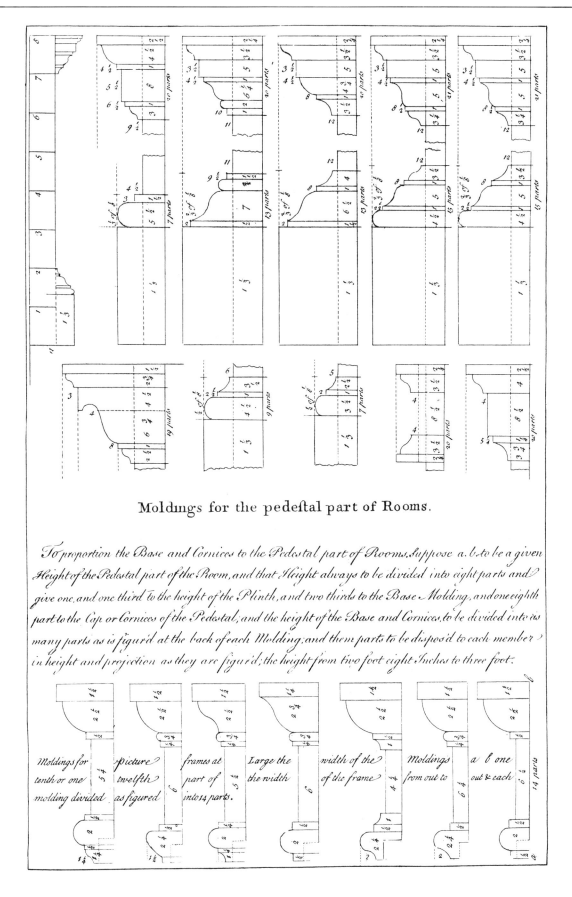

Moldings for the pedeſtal part of Rooms.

To proportion the Base and Cornices to the Pedestal part of Rooms, suppose a.b.to be a given Height of the Pedestal part of the Room, and that Height always to be divided into eight parts and give one, and one third to the height of the Plinth, and two thirds to the Base Molding, and one eighth part to the Cap or Cornices of the Pedestal; and the height of the Base and Cornices, to be divided into as many parts as is figur'd at the back of each Molding; and them parts to be dispos'd to each member in height and projection as they are figur'd; the height from two foot eight Inches to three foot.

Moldings for tenth or one molding divided

picture twelfth as figured

frames at part of into 14 parts.

Large the the width

width of the of the frame

Moldings from out to

a b one out & each

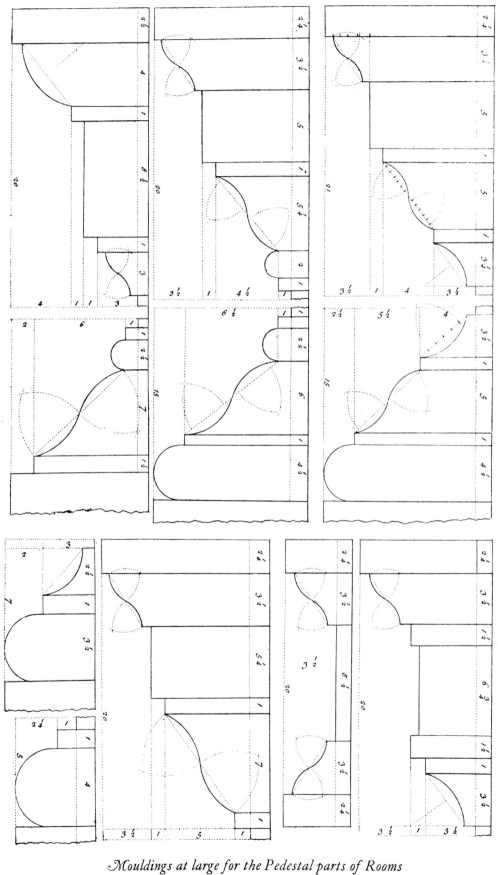

Mouldings at large for the Pedestal parts of Rooms

PROPERTY OF
HIGH POINT PUBLIC LIBRARY
HIGH POINT, NORTH CAROLINA

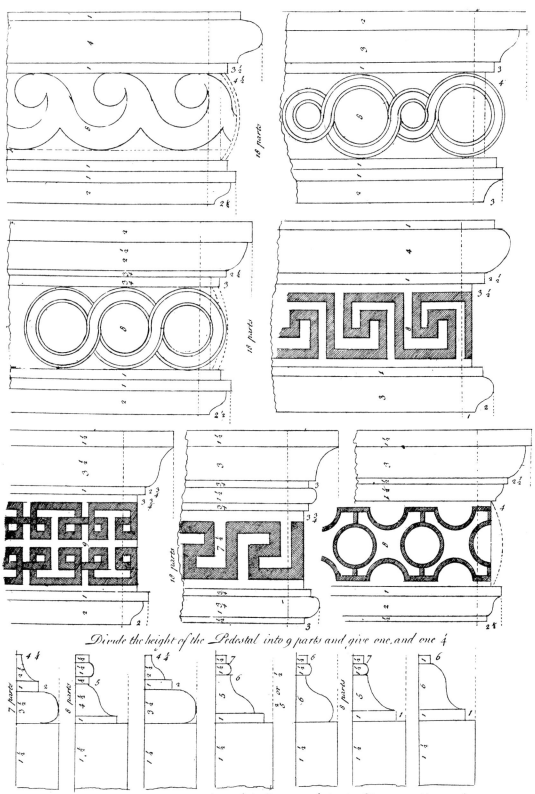

Divide the height of the Pedestal into 9 parts and give one, and one ¼

to the Subbase: 1 and 1½ to the Plinth ⅔ or ½ of the Subbase to the height of the Base-molding: in the open part of the Subbase, the Ornaments are Frets and Guilochis, the heights to be divided into as many parts as they are figur'd, and them parts dispos'd to each member as figur'd.

Base and Subbase for the Pedestal parts of Rooms

Four Chimney pieces with their proportions.

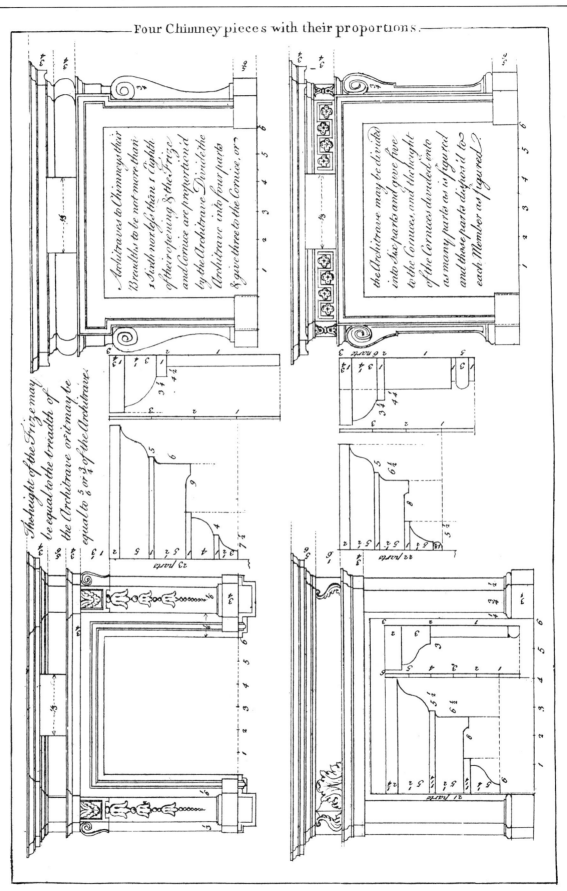

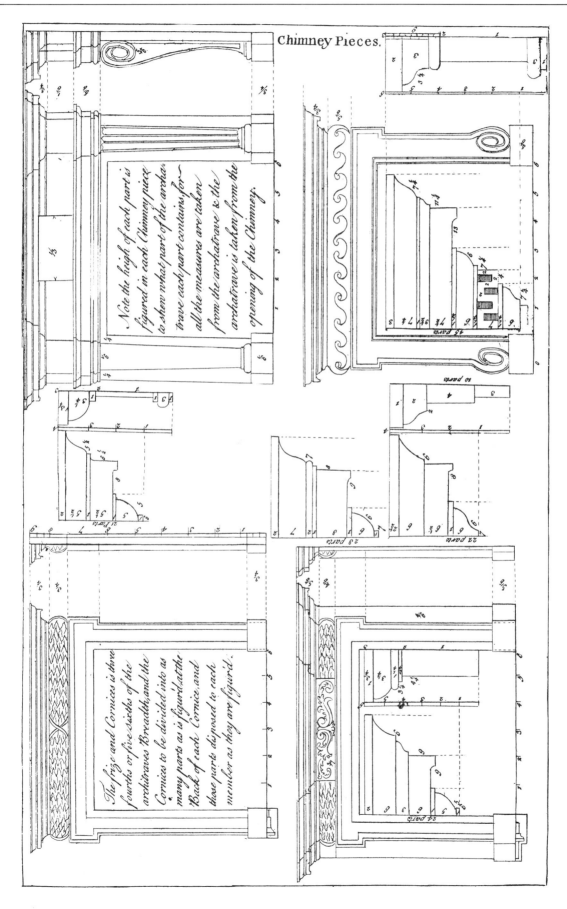

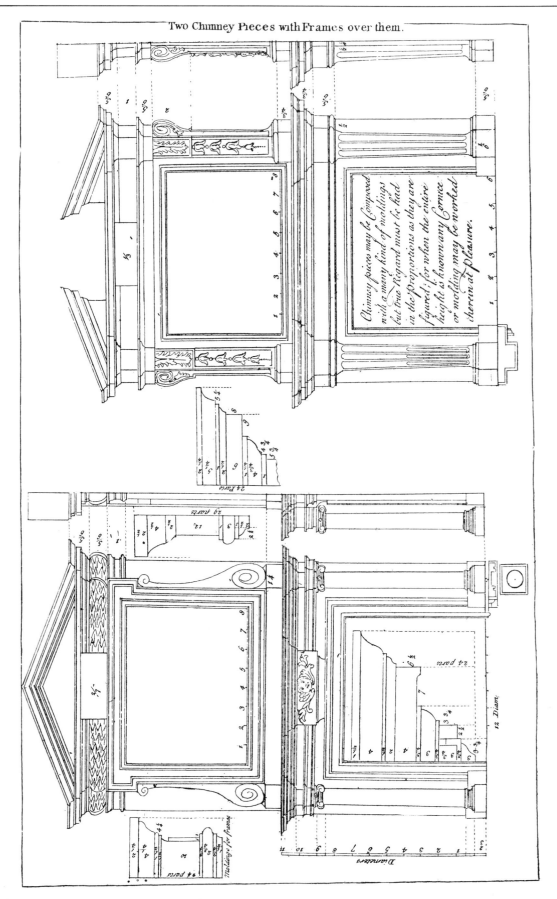

Two Chimney Pieces with Frames over them.

Two Chimney pieces with frames over them.

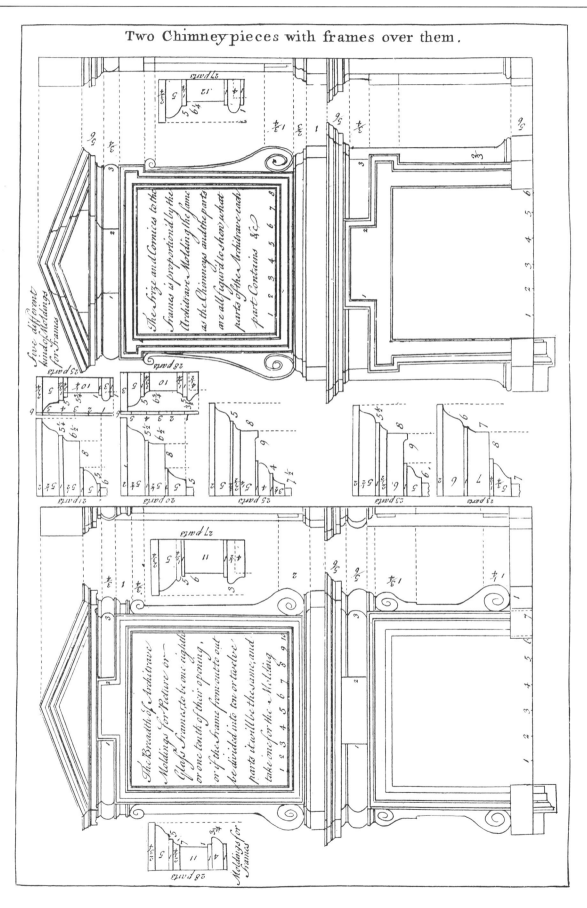

Two Chimney-pieces

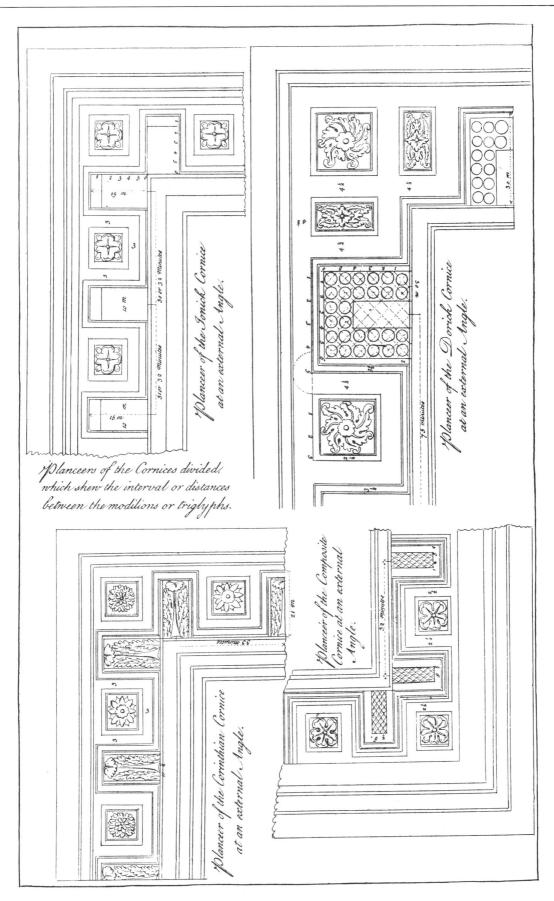

Planceers of the Cornices divided,
which shew the interval or distances
between the modilions or triglyphs.

Plancer of the Ionick Cornice
at an external Angle.

Plancer of the Dorick Cornice
at an external Angle.

Plancer of the Corinthian Cornice
at an external Angle.

Plancer of the Composite
Cornice at an external
Angle.

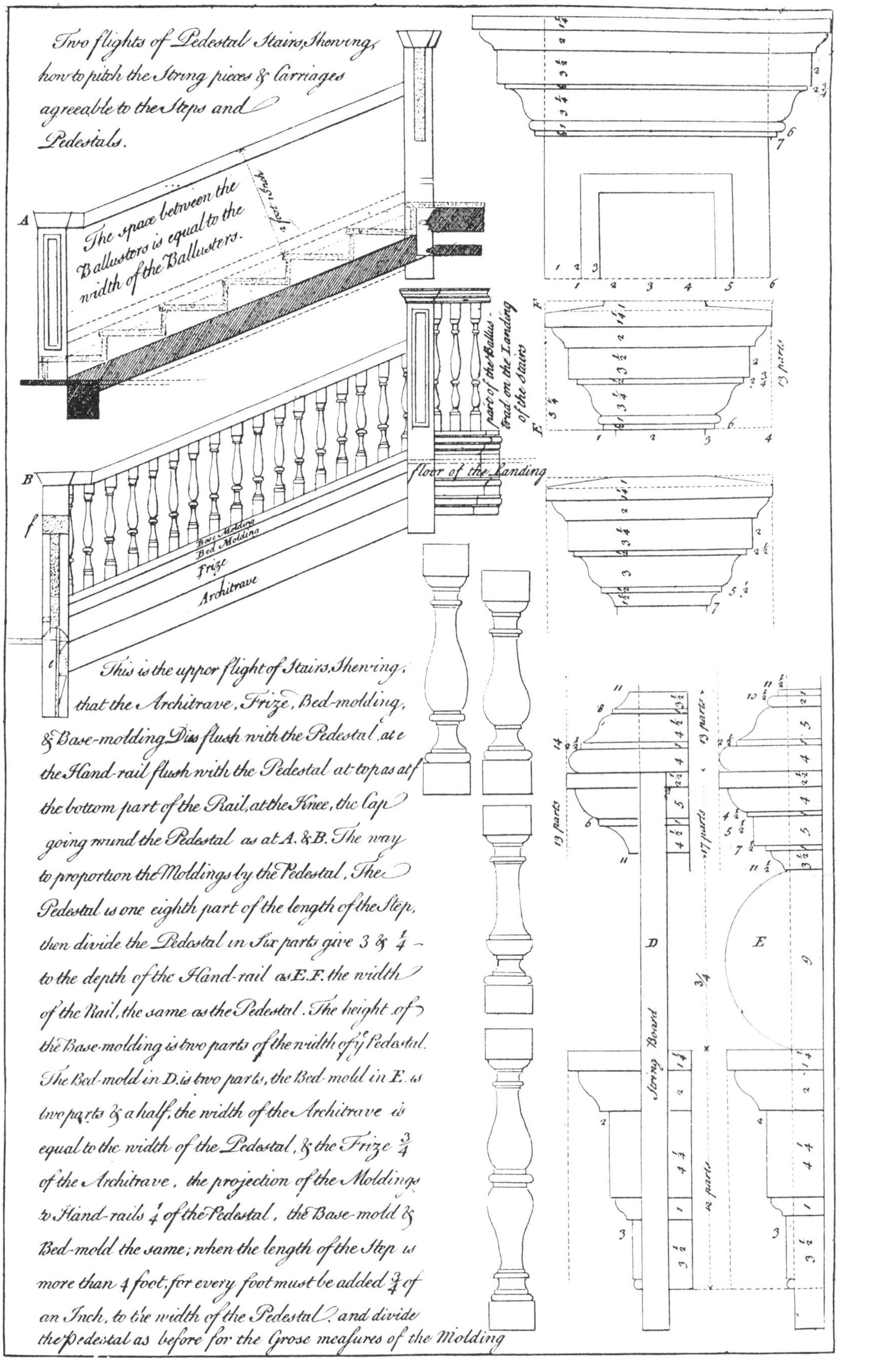

Two flights of Pedestal Stairs, Shewing,
how to pitch the String pieces & Carriages
agreeable to the Steps and
Pedestals.

The space between the
Ballusters is equal to the
width of the Ballusters.

A

pair of the Ballus-
trad on the Landing
of the Stairs

floor of the Landing

B

Box Molding
Bed Molding
Frize
Architrave

This is the upper flight of Stairs, Shewing,
that the Architrave, Frize, Bed-molding,
& Base-molding, Dies flush with the Pedestal, at e
the Hand-rail flush with the Pedestal at top as at f
the bottom part of the Rail, at the Knee, the Cap
going round the Pedestal as at A. & B. The way
to proportion the Moldings by the Pedestal. The
Pedestal is one eighth part of the length of the Step,
then divide the Pedestal in Six parts give 3 & ¼
to the depth of the Hand-rail as E.F. the width
of the Rail, the same as the Pedestal. The height of
the Base-molding is two parts of the width of y Pedestal.
The Bed-mold in D is two parts, the Bed-mold in E is
two parts & a half, the width of the Architrave is
equal to the width of the Pedestal, & the Frize ¾
of the Architrave, the projection of the Moldings
to Hand-rails ¼ of the Pedestal, the Base-mold &
Bed-mold the same; when the length of the Step is
more than 4 foot, for every foot must be added ¾ of
an Inch, to the width of the Pedestal and divide
the pedestal as before for the Grose measures of the Molding

D

String Board

E

F

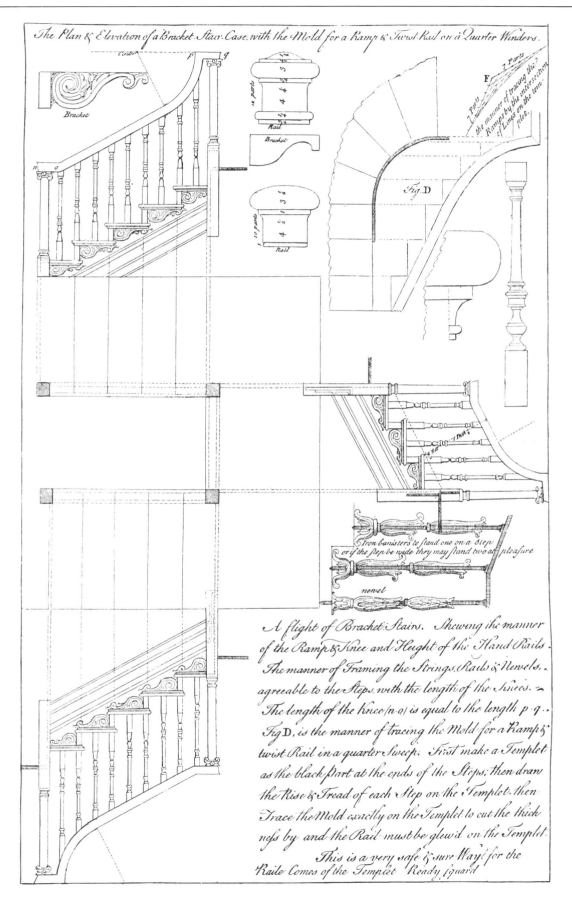

The Plan & Elevation of a Bracket Stair-Case, with the Mold for a Ramp & Twist Rail on a Quarter Winders.

Brackel

Rail

Brackel

Rail

F

7 Parts

7 Parts

7 Parts

the manner of tracing the
Ramps by the intersection
of Lines on the tem-
plet.

Fig. D

Iron banisters to stand one on a step
or if the step be wide they may stand two at pleasure

newel

A flight of Bracket Stairs. Shewing the manner
of the Ramp & Knee and Height of the Hand Rails.
The manner of Framing the Strings, Rails & Newels,
agreeable to the Steps, with the length of the Knees.
The length of the Knee (n o) is equal to the length p q.
Fig D. is the manner of tracing the Mold for a Ramp &
twist Rail in a quarter Sweep. First make a Templet
as the black part at the ends of the Steps; then draw
the Rise & Tread of each Step on the Templet, then
Trace the Mold exactly on the Templet to cut the thick
ness by and the Rail must be glew'd on the Templet.
This is a very safe & sure Way; for the
Raile Comes of the Templet Ready squar'd

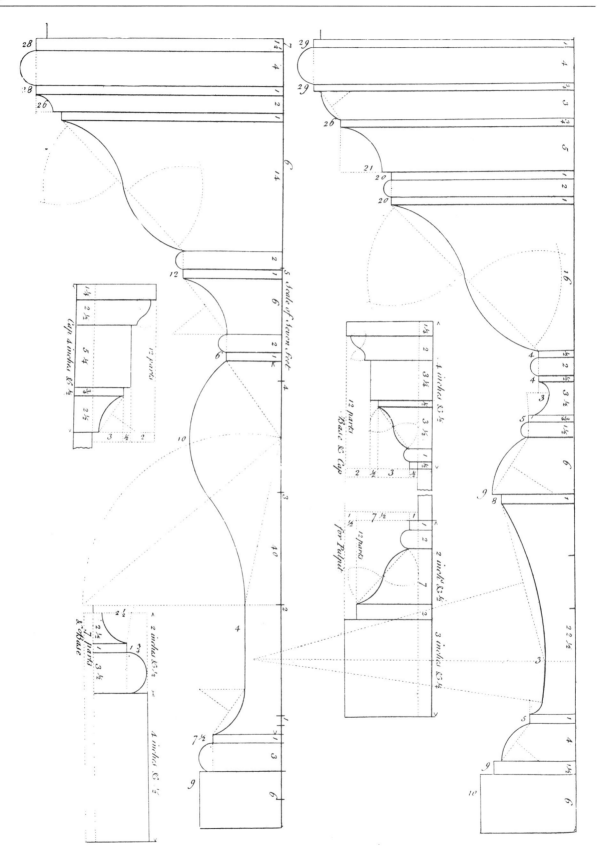

Half of two Supporters for Pulpits at Large,
with the Base and Cornice for the Body of the Pulpits at Large

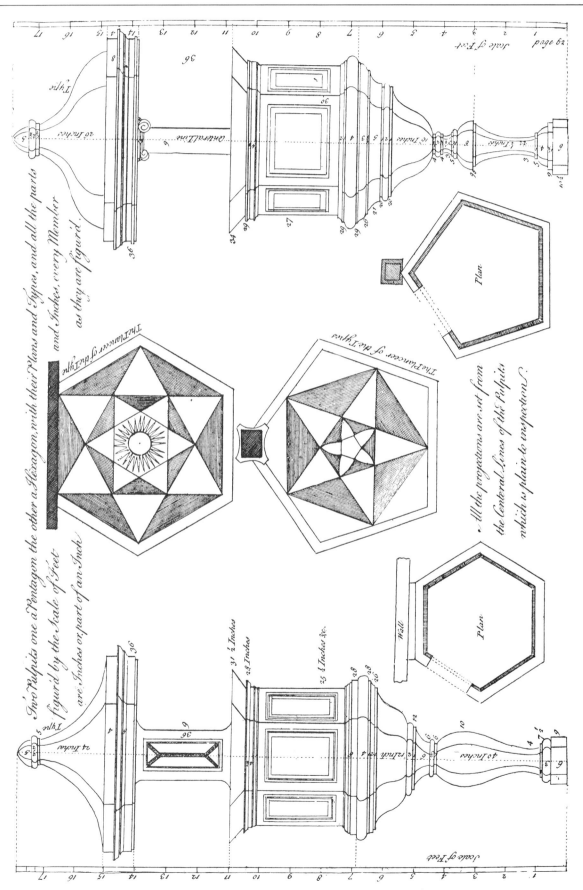

Impost at Large.

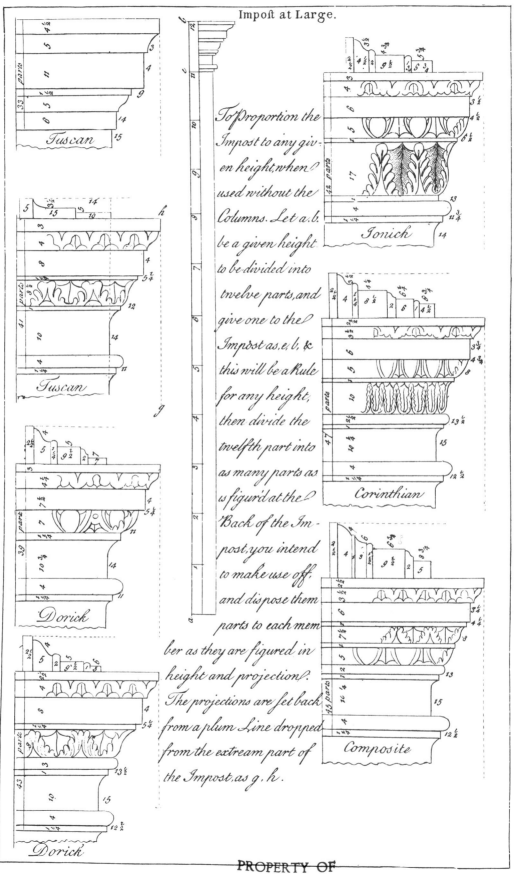

To proportion the Impost to any given height, when used without the Columns. Let a.b. be a given height to be divided into twelve parts, and give one to the Impost as, e, b, & this will be a Rule for any height, then divide the twelfth part into as many parts as is figur'd at the Back of the Impost, you intend to make use off, and dispose them parts to each member as they are figured in height and projection. The projections are set back from a plum Line dropped from the extream part of the Impost, as g. h.

Tuscan

Tuscan

Dorick

Dorick

Ionick

Corinthian

Composite

PROPERTY OF
HIGH POINT PUBLIC LIBRARY
HIGH POINT, NORTH CAROLINA

Early American Roofs

Text by
Aymar Embury II
Photographs by
Kenneth Clark
Originally published in 1932 as White Pine Monograph
Volume XVIII, Number 1

MOUNT VERNON MANSION
The hip roof is especially appropriate to a house of this size.

EARLY AMERICAN ROOFS!

THE roofs are so important a feature of Colonial houses that the various types are often distinguished by names describing their roof shapes, but nobody ever has much to say about them, and few modern houses of Colonial precedent employ anything but the straight gable, or the gambrel roof. Our ancestors were more fluent designers than are we; problems of grade and symmetry which to us are insurmountable seemingly presented no problems to them, and where we content ourselves with the simplest of roof forms they used a multitude of types, sometimes for definite reasons—more often, perhaps, just because they liked them.

It is worthwhile to examine the sketches below just to see how great was the number of their varieties. Amusing names, some of them, calling up the circumstances under which they were built; the common household objects which inspired the forms, or the laws (either enacted or economic) which dictated the methods of roofing adopted by the farmers and sailors who formed so great a proportion of our early housewrights.

There is a word which is not used, but should not be forgotten; we still have shipwrights and wheelwrights by name, but instead of housewrights we have carpenters and cabinet-makers; even "joiners" has come to apply to a collector of societies, rather than

a branch of house carpentry; the legal profession holds its old terms and lawyers are still "Attorneys and Counselors at Law" while the "Carpenter and Joiner" of the 1860's is now only the man behind the sign "Jobbing Done."

We will some day, it may be supposed, give up all our pitched roofs, be they steep or flat or gambrel, and roof our houses with flat slabs of waterproof concrete, or some new processed metal which will not shrink or split as it adapts itself to the hot sun of our long summers or the biting cold of our February nights; but when this occurs, and a pitched roof becomes to our descendants as fantastic as battlements on a stucco cottage, not only will we have lost one of our traditional habits of life, but our northern landscape will have lost the most picturesque accent (next to the church spires) which it possesses.

The house with the flat roof is not necessarily ugly or even unpicturesque; there are plenty of houses in Tunis and Spain and Guatemala to prove that the flat-roofed house may have a charm and beauty all its own; but beneath our northern skies, within our landscape, we must have roofs that show. Take, for example, the little picture at the head of the next page, showing the quarters and shops of Mount Vernon, the small clean white houses with their simple roofs marching along against the foliage of the great trees

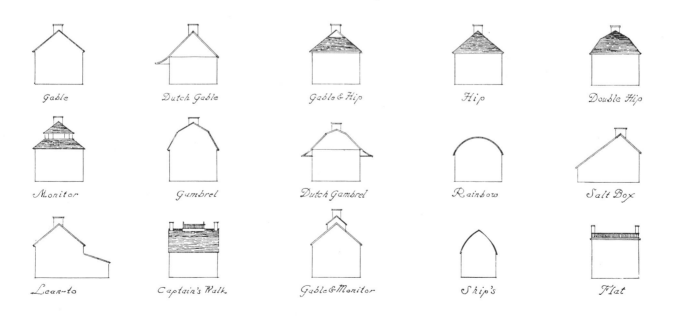

Gable   Dutch Gable   Gable & Hip   Hip   Double Hip

Monitor   Gambrel   Dutch Gambrel   Rainbow   Salt Box

Lean-to   Captain's Walk   Gable & Monitor   Ship's   Flat

OUTBUILDINGS OF MOUNT VERNON MANSION
Simple gable roofs are perfectly suited to these small structures.

full of little holes for the sky to peer through. Not only in this picture, but in reality, they are of breathtaking beauty, not because of any wealth of carving or delicacy of design, but because the simple masses of these early buildings attained once and for all a perfect attunement to our American scene. No one can do a better Parthenon; it is the perfect solution of a simple architectural problem in the Grecian setting, and there is nothing more perfect than perfection. So with our early American work, by some happy accident, or by the expression of obscure instinct, our forefathers achieved in these small white Colonial houses nestling in the shelter of great trees, an absolute rightness which cannot be improved.

Yet, just as in Greece there are other buildings than the Parthenon which are in their ways just as beautiful—the Ionic column is unlike but parallel to the Doric—so in our Colonial houses there were many roof forms, each of which in its proper setting satisfies our aesthetic requirements—and the Colonial designers seem to have felt about them much as we do. On the flat bare plains of Long Island and the wind-swept open seaside dunes, they rarely erected the prim, demurely stately, two-story house of the villages. We find on Long Island the saltbox and the lean-to, in Jersey and along the Down East coast the gambrel or rainbow, types rarely seen in villages, except for the modification of the gambrel used by the Dutch around New York. The hip roof was the hardest to frame and only shows to advantage on buildings of considerable size. Perhaps for these reasons we find it used only on those houses where dignity, or at

least the pretense of it, was desired. The straight gable roof is apt to be over-dominant on the big house, and very likely it was this that caused the main building of Mount Vernon to be built with a hip roof, while the smaller outbuildings have, for the most part, gable ends.

Neither the material of the body of the house nor the part of the country in which it was built appears to have had much influence on the choice of roof design; we find wood, brick, and stone houses with hip roofs and gable roofs; we find hip roofs, gable roofs, and gambrel roofs in New England, around New York, and in the South; apparently the builders in all the colonies knew what was being done in roofs, even if they didn't know how they were built; and there is occasional internal evidence that the builders started a roof of some peculiar form without knowing just how it was to be completed, and finished it by the light of pure reason, rather than by the lamp of experience.

There was, however, a strong local flavor in the design of roofs, just as there was in the choice of scale of ornament; the gambrel of New England was composed of different pitches from that of Maryland, the Pennsylvania gable roof (there much the most popular type) had different relations of height and breadth from those in Massachusetts and Virginia, although, curiously enough, the Pennsylvanians arrived at gable ends of almost exactly the shape common on the eastern end of Long Island; the New England roofs were less steep and the Southern ones steeper. That variance was most likely temperamental, since the greatest difficulty in the way of making a roof tight was snow, and snow is supposed to be more common —one might almost say more prevalent—in New England than in Carolina.

Construction also influenced roof shape, although construction was often more influenced by tradition and desire than by economic factors. In New Jersey for example, the Dutch settlers used much stone; and although stone has been discovered in New England, stone houses have not ("What, never?" "Well, hardly ever!") and although lime was scarce and dear in New Jersey, the Dutchman built of stone just the same; for mortar he used mud. Mud is not hydraulic; so they

NORTON HOUSE—1690—GUILFORD, CONNECTICUT

protected these mud-built houses by wide overhanging roofs, and to get the overhang they swept the eaves out in great curves, producing roof lines of real grace and charm, and almost impossible to ventilate, so that the second stories of these Dutch houses were too hot to sleep in. Papa and Mamma slept on the ground floor, while the children stayed awake on the second.

One style of roof only was peculiar to a single section—the monitor roof to New England. Beginning perhaps as a double-hip roof—a sort of hipped gambrel—it was found easier to make the junction between the two pitches tight

SPENCER-PIERCE HOUSE—1650—NEWBURY, MASSACHUSETTS
The straight gable roof on an unusual house suggestive of an earlier English prototype.

if a vertical board were introduced between them: such a board can be seen on the roof of the Prince House or in a gable-on-hip roof below it. Then perhaps some bright young man bethought himself that if this were raised a foot or two, there would be space for air and light; and behold! the monitor roof. Sometimes this object was accomplished by a super-imposed gable, or by running the ridge of a hipped roof out beyond the structural hip and putting in a window, giving the gable-on-hip roof; sometimes it was flattened and a captain's walk built; and some-times a general amalgamation of monitor, cupola, and captain's walk resulted in a sort of enclosed bridge deck as in the house at Lyme, New Hampshire, a

designed barns and silos built by our progressive farmers. One would have expected the opposite, since our much greater variety of materials makes very simple the problems that must have sorely puzzled our ancestors. The average Colonial house was roofed with shingles; a few in Pennsylvania and the Vermont-New York line with slate; a very few with metal, lead, or at the end of the period, tin-plated iron. Lead, for many years the only material available for flashing, was beaten out by hand with small sheets and was enormously expensive, so that flashing was very sparingly used and in many houses was absent alto-gether. Our modern builders would certainly be puzzled if asked to make a roof tight without it, and

MORRIS-PURDEE HOUSE, MORRIS COVE, NEW HAVEN, CONNECTICUT
In 1670, architects did not feel it necessary to receive minor roofs directly on major ones.

very handsome development, beyond which nothing seemed possible.

Nor have we gone further than they; perhaps, on the contrary, we have retrogressed, if by retrogression we mean a failure to exploit the available possibilities. As was said above we commonly use three forms of roof only, the gable, gambrel, and hip; the more com-plicated forms involving curves, such as the ship, rain-bow, and Dutch, rarely appear, if we except the state-

most carpenters would rebel if asked to shingle a roof with a pitch of 15°, or a rainbow roof, and make it watertight. One trick of the older builders we are beginning to use again, that of canting the ridge lines at the chimneys, was originally a trick to shed water from the chimneys, but is now done to soften roof lines.

Much of our modern "Colonial" work is hard, wiry, correct, and dull; greater variety in roof lines is essential to improvement in this respect.

Rear Porch
GUNSTON HALL—1758—FAIRFAX COUNTY, VIRGINIA
An example of an octagonal hip roof against a wall. A detail drawing on the following page shows it in plan.

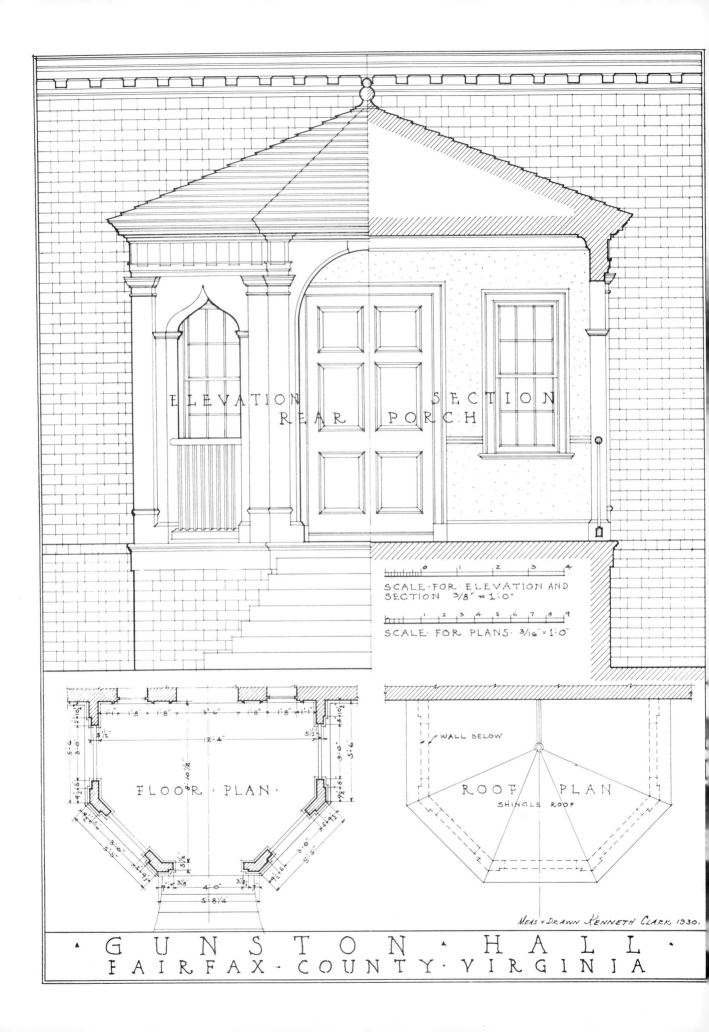

ELEVATION

SECTION

REAR PORCH

SCALE·FOR·ELEVATION·AND
SECTION 3/8" = 1'0"

SCALE·FOR·PLANS· 3/16" = 1'0"

FLOOR·PLAN·

ROOF PLAN
SHINGLE ROOF

WALL BELOW

MEAS & DRAWN KENNETH CLARK 1930.

· G U N S T O N · H A L L ·
F A I R F A X · C O U N T Y · V I R G I N I A

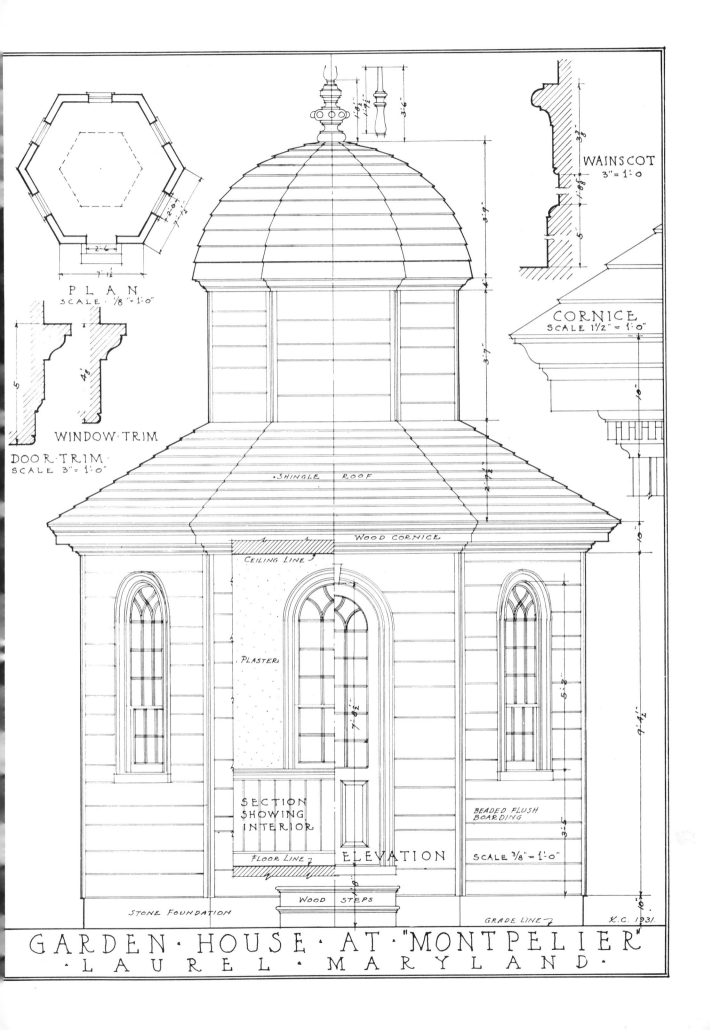

PLAN
SCALE ⅛" = 1'-0"

WINDOW·TRIM

DOOR·TRIM·
SCALE 3" = 1'-0"

WAINSCOT
3" = 1'-0"

CORNICE
SCALE 1½" = 1'-0"

·SHINGLE   ROOF·

WOOD CORNICE

CEILING LINE

PLASTER

SECTION
SHOWING
INTERIOR

FLOOR LINE

BEADED FLUSH
BOARDING

SCALE ⅜" = 1'-0"

ELEVATION

STONE FOUNDATION

WOOD   STEPS

GRADE LINE

X.C. 1931.

GARDEN·HOUSE·AT·"MONTPELIER"
·LAUREL·MARYLAND·

GARDEN HOUSE, MONTPELIER, LAUREL, MARYLAND
A hexagonal dome over a hexagonal hip roof.

DYCKMAN HOUSE—1787—NEW YORK, NEW YORK

JOHN P. B. WESTERVELT HOUSE, CRESKILL, NEW JERSEY

Main house built with mud mortar, needing protection on gable
end. Addition built with lime mortar and gable end unprotected.

CAPTAIN JOHN CLARK HOUSE, SOUTH CANTERBURY, CONNECTICUT
Built in 1732, enlarged about 1790. Gable on hip roof to receive chimney.

CHAMPION HOUSE—1794—EAST HADDAM, CONNECTICUT
A hip roof so flat that a balustrade appears to be necessary.

GOVERNOR TRUMBULL HOUSE—1753—LEBANON, CONNECTICUT
A hip roof with short ridge.

WITTER HOUSE—1828—CHAPLIN, CONNECTICUT
Early type of monitor used on house of later date.

A HOUSE NEAR WESTMORELAND, NEW HAMPSHIRE
A hip roof with a lean-to which was not a later addition but part of the original structure.

HOUSE AT LYME, NEW HAMPSHIRE
A late and handsome development of monitor hip roof.

PRINCE HOUSE, FLUSHING, NEW YORK
A gambrel and gable roof used on a late eighteenth-century house.

A 1757 HOUSE, EAST GREENWICH, RHODE ISLAND
Another example of the gable on hip roof—this time with a window.

PROPERTY OF
HIGH POINT PUBLIC LIBRARY
HIGH POINT, NORTH CAROLINA